Dear Rob,

I hope you enjoy the book and look forward to hearing your thoughts.

Best wishes,

Tim

D1789316

# The Politics of Nuclear Disarmament

This book explores what political conditions must be established and what obstacles overcome for the five official Nuclear Weapon States (NWS)—China, France, Russia, the UK and US—to eliminate their nuclear weapons.

The different views and positions of a range of actors concerning nuclear weapons issues—including elite perspectives and public opinion—and the political assumptions underpinning them, are discussed to develop a more democratic approach to disarmament. Addressing the lack of detailed analysis concerning the meaning of nuclear disarmament for the domestic political orders of NWS, the book critically explores different approaches to and theories of disarmament within legal, political and technical literatures and orthodox and critical theory. It also builds on previous discussions of nuclear possession, restraint, arms control, and disarmament—concerning both nuclear possessor and non-possessor states—identifying the insights these works provide regarding how NWS disarmament may be advanced.

Contributing to theoretical debates concerning how domestic politics interacts with and determines states' international behaviour, the book will be of interest to all scholars and students of history, politics, international relations, security studies, military history, war studies, peace studies, conflict, democracy, and global governance.

**Tim Street** is a researcher and activist based in Chelmsford and Oxford.

**Modern Security Studies**
Series editors: Sean S. Costigan and Kenneth W. Estes

This series fills a known gap in modern security studies literature by pursuing a curated, forward-looking editorial approach on looming and evergreen security challenges. Short and long form works will be considered with an eye towards developing content that is widely suitable for instruction and research alike. Works adhere around the series' four main categories: Controversies, Cases, Trends and Primers. We invite proposals that pay particular attention to controversies in international security, notably those that have resulted in newly exposed and poorly defined risks to non-state legitimacy, international or state capacities to act, and shifts in global governance. Case studies should examine recent historical events and security-related actions that have altered present day understanding or political calculations. Trends will need to detail future yet tangible concerns in a five to ten year timeframe. Authors are also invited to submit proposals to our primers category for short form works on key topics that are referenced and taught throughout security studies.

For more information about this series, please visit: www.routledge.com/politics/series/ASHSER1437

**Verifying Nuclear Disarmament**
*Thomas Shea*

**NATO's Democratic Retrenchment**
Hegemony after the Return of History
*Henrik B. L. Larsen*

**The Politics and Technology of Cyberspace**
*Danny Steed*

**Post-Cold War Anglo-American Military Intervention**
A Study of the Dynamics of Legality and Legitimacy
*James F. D. Fiddes*

**Nuclear Modernization in the 21st Century**
*Edited by Aiden Warren and Philip M. Baxter*

**The Politics of Nuclear Disarmament**
Obstacles to and Opportunities for Eliminating Nuclear Weapons
*Tim Street*

# The Politics of Nuclear Disarmament

Obstacles to and Opportunities for Eliminating Nuclear Weapons

**Tim Street**

LONDON AND NEW YORK

First published 2021
by Routledge
2 Park Square, Milton Park, Abingdon, Oxon OX14 4RN

and by Routledge
52 Vanderbilt Avenue, New York, NY 10017

*Routledge is an imprint of the Taylor & Francis Group, an informa business*

*British Library Cataloguing-in-Publication Data*
A catalogue record for this book is available from the British Library

*Library of Congress Cataloging-in-Publication Data*
A catalog record has been requested for this book

ISBN: 978-0-367-49129-1 (hbk)
ISBN: 978-0-367-74376-5 (pbk)
ISBN: 978-1-00-304469-7 (ebk)

Typeset in Times New Roman
by Newgen Publishing UK

# ·Contents

# About the author

**Tim Street** is a researcher and activist based in Oxford. Tim has worked on peace and disarmament issues since 2005 with a range of groups including Campaign Against Arms Trade, the International Campaign to Abolish Nuclear Weapons, British American Security Information Council, Conscience and Drone Wars UK. Tim recently completed his PhD exploring the politics of nuclear disarmament at Warwick University. He is an Associate Fellow with Oxford Research Group and a board member of Nuclear Information Service.

# Note on terminology

The 'official' nuclear weapon states (NWS) under the nuclear non-proliferation treaty (NPT) are China, France, Russia, the UK and US. The NPT defines an NWS as a state 'which has manufactured and exploded a nuclear weapon or other nuclear explosive device prior to 1 January 1967'.[1] Member states of the NPT without nuclear weapons are known as non-nuclear weapon states (NNWS). The designation of former nuclear weapon state (FNWS) is used to refer to NWS that in future complete the disarmament process, for the political and technical reasons explained below. The four nuclear-armed states (NAS) that are not members of the NPT are Israel, North Korea, India and Pakistan. The NWS and NAS are collectively referred to as nuclear weapon possessor states (NPS).

## Note

1  United Nations (1968), *Treaty on the Non-Proliferation of Nuclear Weapons*, www.un.org

# Abbreviations

| | |
|---|---|
| A2/AD | Anti-access and area denial |
| ABMT | Anti-ballistic missile treaty |
| AI | Artificial intelligence |
| AWE | Atomic Weapons Establishment |
| AWS | Autonomous weapon systems |
| BASIC | British American Security Information Council |
| BMD | Ballistic missile defense |
| CCP | Chinese Communist Party |
| CEA | Commissariat à l'énergie atomique |
| CFE | Treaty on Conventional Armed Forces in Europe |
| CND | Campaign for Nuclear Disarmament |
| CPGS | Conventional prompt global strike |
| CTBT | Comprehensive test ban treaty |
| DOD | Department of Defense |
| DOE | Department of Environment |
| EU | European Union |
| FMCT | Fissile material cut-off treaty |
| FNWS | Former nuclear weapon state |
| HEU | Highly enriched uranium |
| ICBM | Intercontinental ballistic missile |
| ICJ | International Court of Justice |
| IMEMO | Russian Institute of World Economy and International Relations |
| INF | Intermediate-range nuclear forces treaty |
| MAD | Mutual assured destruction |
| MIC | Military industrial complex |
| MOD | Ministry of Defence |
| NAM | Non-aligned movement |
| NAS | Nuclear armed state |
| NATO | North Atlantic Treaty Organisation |
| NC3 | Nuclear command and control |
| NGO | Non-governmental organisation |
| NNSA | National Nuclear Security Administration |

| | |
|---|---|
| NNWS | Non-nuclear weapon state |
| NPG | Nuclear planning group |
| NPR | Nuclear posture review |
| NPS | Nuclear possessor state |
| NPT | Nuclear non-proliferation treaty |
| NSS | National security strategy |
| NWFW | Nuclear weapons free world |
| NWS | Nuclear weapon state |
| OECD | Organisation for Economic Co-operation and Development |
| PLA | People's Liberation Army |
| PNND | Parliamentarians for Nuclear Non-Proliferation and Disarmament |
| PRC | People's Republic of China |
| RevCon | Review conference |
| SIPRI | Stockholm International Peace Research Institute |
| SNP | Scottish National Party |
| SSBN | Ship submersible ballistic nuclear |
| START | Strategic Arms Reduction Treaty |
| TNW | Tactical nuclear weapons |
| TPNW | Treaty on the prohibition of nuclear weapons |
| UK | United Kingdom |
| UN | United Nations |
| US | United States of America |
| USSR | Union of Soviet Socialist Republics |
| WMD | Weapons of mass destruction |
| WW2 | World War Two |

# Introduction

## I.1 Why nuclear disarmament?

Despite humanity having accumulated a range of relevant knowledge and experience, eliminating nuclear weapons—the most powerful weapons ever invented—will, for the nuclear possessor states (NPS) and the rest of the world, essentially be a voyage into the unknown. For those who believe that nuclear weapons bring security, stability and freedom to their nations, including prominent figures such as Malcolm Rifkind, Brad Roberts and Bruno Tertrais, this voyage is unappealing because it is fraught with costs, risks and uncertainty given the high stakes and limited benefits involved.[1] Other opponents of disarmament, such as former senior White House officials Harold Brown and John Deutch, also claim that the nuclear 'genie' cannot be put back in the bottle and that nuclear weapons cannot be disinvented.[2]

Yet for those who reject nuclear weapons as illegal, immoral and insane—including prominent voices in the global nuclear disarmament movement such as Desmond Tutu, David Krieger and the Women's International League for Peace and Freedom—this voyage is essential if humanity is to reach safe haven and liberate itself from the dominance of the overly powerful and the continued threat of annihilation.[3] From this perspective, the risks involved in disarming are manageable and ultimately negligible compared to those of the status quo. The nuclear genie can thus be dealt with by delegitimising the bomb so that it is as socially unacceptable as cannibalism or slavery.[4] In addition, sensitive nuclear weapons information may be destroyed and related knowledge and skills may be allowed to atrophy.[5] Whichever position one takes on this issue, mapping out as far as possible the terrain that will likely need to be traversed in order to move towards abolition is a useful task if governments and people everywhere are to make informed decisions regarding the future of nuclear weapons—both for their own nations and the world.

As we shall see, the idea and imagery of time are as important to this discussion as that of space, given that nuclear disarmament is a social and political process without a final end point. For Jonathan Schell, if a nuclear weapons free world (NWFW) is to be created, this therefore requires a political solution that aims at being 'global and everlasting'.[6] According to this speculative logic, even if the voyage of discovery for each NPS leads to them eliminating their nuclear weapons, this zero state will need to be maintained through good relations between nations, anchored in durable institutions, given that, as Ian Anthony points out, 'the knowledge needed to rebuild nuclear weapons will never be forgotten, and by extension there is no exclusively technical guarantee against nuclear re-armament'.[7]

Time is also important in the sense that nuclear disarmament may be seen as an urgent necessity because of the significant risk of nuclear detonation that would likely, in most scenarios, have catastrophic consequences. Seth Baum has therefore argued that 'Nuclear war is the black swan we can never see, except in that brief moment when it is killing us. We delay eliminating the risk at our own peril.'[8] Concern over nuclear proliferation and rising international tensions led the *Bulletin of the Atomic Scientists* to announce in 2020 that the world is '100 seconds to midnight' because:

> Humanity continues to face two simultaneous existential dangers—nuclear war and climate change—that are compounded by a threat multiplier, cyber-enabled information warfare, that undercuts society's ability to respond. The international security situation is dire, not just because these threats exist, but because world leaders have allowed the international political infrastructure for managing them to erode.[9]

The continuing threat posed to humanity by nuclear weapons and the instability and fallibility of deterrence lends significant weight to the argument that nuclear disarmament is vital, as a global public good. The need for fresh and clear thinking on this issue is also important because, as the Weapons of Mass Destruction Commission notes, despite hopes of a peace dividend following the end of the Cold War, disarmament is today 'in disarray' as the US pursued 'absolute global superiority' and denigrated 'international institutions and instruments'.[10] Furthermore, as Catherine Kelleher highlights, there has been an 'unknowing' of previous disarmament efforts so that a new generation of officials and students are unfamiliar with the relevant history.[11] Similarly, Lawrence Wittner identifies the 1980s as the 'heyday' of the nuclear disarmament movement and argues that today's movement is 'considerably weaker'.[12] Based on these salutary observations, it is clear that academia, civil society and the governments of NPS need to devote far more resources towards creating the time and space for disarmament if it is to advance.

Those who adopt a normative position in favour of nuclear disarmament may be said to follow natural historians such as Peter Kropotkin, taking an optimistic view of human nature and the potential for people to use their intelligence, creativity and skills for cooperative and peaceful, rather than hostile and warlike ends, as suggested by social Darwinism.[13] Yet, as Bertrand Russell argued, the 'division of man into competing and often hostile nations' and the emergence of 'scientific man' capable of creating nuclear weapons has led to humanity facing the question—'Shall we put an end to the human race; or shall mankind renounce war?'[14] Russell's observations may lead us to consider not only who and what stands in the way of more equitable and just societies but who and what stands in the way of the prospects for decent survival on this planet.

This study addresses these questions by seeking to understand what the existence, use and abolition of nuclear weapons means for the world in relation to the principles of equality and justice. It is argued that these issues can only be dealt with through an assessment of each state's strategic power and behaviour. To begin with, there is a much greater concentration of such power in NPS, held in the hands of small elite groups. To properly map these power structures and differentiate the types and degree of power held between different actors it is necessary to ask several further questions. For example, which people, groups and institutions legitimate, value and reproduce nuclear weapons and resist demands for disarmament? Why do they do this? What impact do nuclear weapons have on domestic and international politics and what drives nuclear proliferation? And what is the current state of the national and international movements for nuclear abolition?

In seeking to answer these questions, this study has itself been bound by limits of time and space that raise certain practical and theoretical problems. For example, whilst there are nine NPS and 180 non-nuclear weapon states (NNWS), this study principally focuses on the politics of nuclear disarmament in the five nuclear weapon states (NWS), so that neither the four nuclear armed states (NAS), nor the NNWS, are addressed in any depth. In order for the politics of a NWFW to be fully explored, the obstacles to and opportunities for disarmament in and between all the NPS and NNWS would need to be covered. The intention here is thus to contribute to the existing literature and lay some of the groundwork for such future studies. In any case, it is reasonable to focus solely on the NWS as a grouping given their obligations to disarm under the NPT, the fact that Russia and the US possess approximately 91% of the world's 13,410 nuclear warheads, and the immense literature on nuclear matters which this study has attempted to select from and discuss as judiciously as possible.[15]

Even for the limited scope of this study, the exclusion of the NAS and NNWS is problematic given that moves towards disarmament involving the NWS will, at some stage, also have to consider how NAS and NNWS can and should be involved. This again raises the question of where responsibility for NWS disarmament lies and how it may be differentiated between relevant actors both internal and external to NWS. The approach I have taken to this issue is to highlight, where appropriate, the relationships between NWS, NAS and NNWS to give a sense of the role nuclear weapons play in shaping the wider international order. I do this primarily in Part II when discussing nuclear politics involving each of the NWS. In doing so, I argue that NWS have a responsibility for both national nuclear disarmament and to act in ways that support the creation of a NWFW. Moreover, I posit that the US has a particularly great responsibility here given its immense military might, which has a singular impact on all other nation's strategic thought.

## I.2  Situating this study in the literature

Whilst there is a wide range of works from academic and non-academic authors addressing the topic of abolishing nuclear weapons, many just focus on juridical and/or technical aspects, excluding substantive political discussions on a domestic and international level. Examples of the former include studies regarding existing and possible legal instruments supportive of disarmament, whilst studies in the latter category include discussions of the different levels of irreversibility or verification requirements involved in disarmament. Furthermore, many of those taking a political approach to disarmament tend to limit themselves for one reason or another. This can occur, for example, when studies are partisan and do not recognise their underlying assumptions, which can lead to a narrow and Western-centric analysis. Other studies suffer by omitting important types of information, whether legal, technical or political. In the case of several mainstream studies this includes the part domestic politics—for example, public opinion, the role and nature of the state and political economy—has and might play in nuclear weapons decision-making, past, present and future.

It will be argued that this state of affairs has often led academics, activists, experts and government officials to talk past one another when discussing nuclear matters, because their approach to the subject is based on different worldviews, so that they use and understand ideas and language around nuclear disarmament in quite different ways. The result is that there is no shared grammar, or story, concerning what nuclear disarmament is and might mean in practice. We can speculate on the causes of this problem—the lack of public awareness, engagement and understanding of nuclear issues— particularly after the Cold War, the fact nuclear disarmament is in several aspects 'unknowable', or the way in which nuclear disarmament is seen as taboo for some given its political implications, thus becoming a site of ideological contestation.

To get a general sense of how nuclear disarmament has been conceived it is useful to sketch out and compare two of the more prominent current schools of thought on the issue. For example, a view held amongst many pro-disarmament activists and campaigners is that banning the bomb is mainly, if not all, about getting rid of nuclear arsenals and that all such weapons are equally abhorrent.[16] According to this universalist view, which is primarily driven by a moral and humanitarian impulse, using legal instruments to ban nuclear weapons would not mean significant changes to the current international order, would be relatively straightforward in terms of costs and technical requirements and should be isolated from debates about national and regional security to focus on common human needs. Many within the disarmament movement, including those from more radical traditions, also highlight the domestic political impact of nuclear possession, particularly the undemocratic and secretive nature of nuclear weapon systems, and argue that disarmament and demilitarisation is necessary for social justice at home and abroad.

A different view, held typically by academics, government officials and policy professionals working within mainstream bodies in both East and West, is that nuclear deterrence is highly valuable as a means of preventing war between the great powers and ensuring international stability. This is because the revolutionary nature of nuclear weapons ensures that no matter how powerful a nation's military is, these weapons are impossible to defend against. A nuclear possessor may therefore prevent an attack by an otherwise stronger aggressor by threatening a nuclear response. The issue of nuclear disarmament is thus bound up with military and state power more broadly, since conventional and nuclear weapons have important qualitative differences with strategic consequences.[17]

According to this view, which is primarily driven by the political and security interests of power elites, a NWFW would entail a transformed international order, would likely be costly, destabilising and technically complex, and cannot therefore be isolated from debates about security. Moreover, this school of thought largely marginalises or ignores the domestic political causes and consequences of nuclear possession and disarmament. Nuclear deterrence is often also presented as being an unfortunate necessity to ensure national survival in a dangerous and uncertain world, with disarmament a noble but distant ambition. Notably, such thinking is as often found amongst elite circles in NPS with formally liberal and democratic regimes as it is in authoritarian states.

These brief sketches—of what might be termed idealist and realist approaches—whilst simplified, illustrate important truths about the different ways nuclear disarmament is seen. Yet there is little commerce or interaction between these two schools of thought, which tend to exist in their own institutions and circuits. The reason such an engagement would be important is that each school contains gaps in its evidential base, and thus its reasoning, that tend to limit the veracity of its conclusions regarding the causes and consequences of disarmament. Moreover, there are aspects of both schools that are complementary and could be brought together to produce an approach to disarmament including, as Stephen Cimbala puts it, 'a compelling moral imperative and a policy prescription that is strategically sound'.[18] As I shall discuss further below, for the purposes of this study I therefore adapt and update ideas from the disarmament movement, particularly its 1980s heyday, which take a more radical view of the nuclear disarmament debate, and apply their approach to the technologically advanced strategic environment of today. A key argument of this more radical view is that, if nuclear disarmament is to advance, democratic and popular movements need to become sufficiently capable of influencing or controlling state institutions so that power structures supporting and reproducing nuclear weapons can be dismantled.

## I.3  Contributions to the literature

In order to address the gaps in the literature identified above, this study has sought to cast a wide net to discover and question the different political assumptions and positions of those engaged in nuclear weapons issues, to create an approach which draws on legal and technical knowledge and integrates idealist and realist visions of disarmament. In conducting the research for this study, both through studying official and non-governmental publications and interviewing high-profile academics, campaigners and officials, it became clear that it was necessary to create such a debate and conceptual synthesis for several reasons. First, as noted above, those currently working on these issues—including some of the most knowledgeable and committed practitioners—are often not familiar or do not engage substantively with arguments and perspectives from sources outside their professional and social circle, including from other countries.

Secondly, whilst there have, episodically, been bursts of studies published presenting action plans for NPS to eliminate their nuclear arsenals, for example, following the arrival in office of Barack Obama as US President in 2009, such works (including lengthy reports by the Canberra Commission, the Middle Powers Initiative, the International Commission on Nuclear Non-Proliferation and Disarmament, and Global Zero—to name but a few) have also suffered by excluding full analyses of the political meaning of NWS disarmament.[19] Despite the many valuable insights regarding the necessary steps and processes involved in disarmament, there was a tendency for these works to significantly underappreciate the transformative nature of abolition for domestic, regional and international power structures, with the sources and scale of the requisite political movements for reform side-lined or ignored.

Moreover, whilst the Stimson Center's *Unblocking the Road to Zero* project provided accessible and detailed political analyses of the NPS and nuclear capable nations, the perspective of citizens and the more transformative vision of the disarmament movement was again absent.[20] The national analyses were also presented separately rather than being brought together to consider the structural implications of nuclear weapons for international order. This study therefore brings together a range of data concerning public attitudes to nuclear weapons and abolition as well as economic and foreign policy—from across NWS—to fashion a more democratic way of thinking about disarmament. Ultimately, this necessitates a focus on the domestic political causes and consequences of NWS acquiring and developing the bomb. The corollary of this proposition is that nuclear disarmament requires a theory of domestic political change, based on the need for popular and democratic control of state institutions as a prerequisite of nuclear disarmament, an idea that I refer to as *institutional democratisation*.

Thirdly, this study aims to inform the wider political and international relations literature by emphasising the importance of nuclear matters to today's global political order. For example, a reconsideration of both the meaning of nuclear weapons and disarmament can enrich and shed new light on key debates, particularly concerning the nature of the modern state, conflict, measurements of democracy, as well as interstate cooperation. The world is facing a series of interlocking and formidable challenges, including climate change, conflict, resource depletion, hunger, poverty and terrorism, so that new insights—such as this study aims to provide—concerning how to manage and ultimately dispense with the power of nuclear arsenals and develop a democratic, just and sustainable power transition towards a NWFW will have implications across subject areas and disciplines.

Fourthly, through investigating the domestic political drivers of nuclear possession and disarmament, this study aims to contribute to theoretical debates concerning how politics at the domestic level interacts with and determines states' international behaviour. Prominent critics of realist theory, such as Phillip Gourevitch and Ethan Kapstein, have highlighted the need for greater discussion and research in this area. Such authors argue that the limitations of realism, for example, concerning the agent-structure problem and its explanation of political change, requires further study, in Kapstein's words, to 'articulate an explicit model of how a given set of domestic factors can produce particular international outcomes, the most important being war and peace'.[21]

## I.4 Study overview

The study consists of eight chapters spread over two parts. Part I comprises Chapters 1 and 2 whilst Part II comprises Chapters 3 to 8. Overall, the first two chapters are focused on providing an introduction to the main problems of nuclear disarmament, discussing both its meaning and significance and the different ways it has been conceptualised, allowing the study to be situated within the literature on nuclear weapons and disarmament as well as political theory. This is done to enable the development of an appropriate methodology to examine the domestic *and* international politics of nuclear weapons concerning NWS.

More specifically, Chapter 1 begins by investigating the meaning and implications of nuclear disarmament, surveying key works in order to consider how it has been used in theory and practice from a legal, political and technical viewpoint. In particular, the importance of international cooperation is highlighted, both in terms of easing tensions to create a suitable environment for disarmament and in terms of developing the monitoring and verification regime needed to support the phases of restraint and elimination on the path to zero. This discussion lays the foundation for the definition and theory of nuclear disarmament that is developed over the course of this study. For example, from a technical point of view, nuclear disarmament is an adaptive process which may be modified to suit situational requirements. Furthermore, different levels of irreversibility for disarmament require varying levels of time and resource commitments. Such insights help us identify relevant sources and forms of data to better imagine the political conditions for, and shape of, alternative non-nuclear futures according to the predicament of each NWS.

As noted above, there are powerful objections to the elimination of nuclear arsenals, for example, regarding the stability of regional and international relations following disarmament and the security dilemmas that, some contend, this would generate. These concerns, it is argued, present formidable barriers to the NWS disarming and, in particular, a NWFW. Given the weight and prominence of these arguments in mainstream thought it is appropriate for a study of this nature, which seeks to examine the politics of nuclear disarmament from a critical and normative perspective, to fully engage with the established literature to provide a rounded and nuanced investigation.

In order to address the arguments presented by works with a more orthodox bent, Chapter 2 discusses existing approaches to nuclear possession and disarmament in realist thought and scholarly critiques of realism, with three main purposes in mind. The first is to assess the extent to which such works provide useful insights into the conditions and indicators of nuclear disarmament, which may be applied to the problem of how to eliminate nuclear weapons in and between NWS. The second is to address what several scholars see as the major issue of how international anarchy spreads fear and uncertainty, which drives conflict and stifles interstate cooperation—which is vital if disarmament is to progress. The third is, where necessary and useful, to improve on existing approaches and positions by developing a new theory that can better explain and respond to the challenges of achieving nuclear disarmament in and between NWS.

For example, having identified the key gaps in and problems with the mainstream literature, I propose a new approach, which I term *institutional democratisation*. This approach, which is based on a normative stance supportive of nuclear disarmament action, draws on insights from global governance and civil society literatures as well as social and political theory. *Institutional democratisation* can be summarised as the idea that the democratic deficit in the five NWS—including, crucially, the US, given its singular influence, power and global reach—needs to be addressed if there is to be any prospect of the NWS embracing nuclear disarmament.

In order to elaborate on the previous discussion of the importance of *institutional democratisation* and the practicalities of abolition, Chapters 3 to 7 present a series of in-depth analyses of the nuclear weapons systems of each of the five NWS in order of their acquiring the bomb: US, Russia, UK, France, China. This includes an assessment of the economic, social, political and technological meanings, ideological underpinnings and institutions that make possible the production and deployment of nuclear weapons, in order to understand both the significance and value assigned to the bomb by elites and the wider society in each NWS, and how the elimination of nuclear arsenals may be accomplished.

In doing so, I emphasise the concentration of nuclear weapons decision-making power, in each NWS, in the hands of a very small number of military and political officials. Moreover, nuclear weapons systems have their own bureaucratic and technological momentum so that their modernisation has often occurred outside of full political control. It is therefore argued that the high secrecy surrounding the development and reproduction of nuclear weapons, along with the highly centralised decision-making structures regarding their use, is incompatible with and corrosive to the spirit and functioning of democracy. With the possible exception of China, primarily given its history of nuclear restraint, authoritarian political system and its perceived need to deter threats from the US, nuclear disarmament for each NWS will therefore likely require, and certainly benefit from, a domestic power transition involving citizens forming popular movements to gain control of state institutions that can manage, restrain and ultimately dismantle nuclear weapons systems. *Institutional democratisation*, as a driver of disarmament, will likely also benefit from other developments that reduce the salience of nuclear weapons, such as a wider societal awareness of these weapon's economic and social costs, and technological advancements that may reduce these weapon's credibility and utility.

In terms of methodology, the investigation presented in Part II of prominent beliefs and perspectives on nuclear possession and disarmament in NWS is based on a critical analysis of academic, advocacy, media and government documents, primarily from sources based in the five NWS, which variously include: historical works; policy and briefing papers; official statements; media reports; public opinion polls; speeches and debates. This analysis was enhanced by insights gathered from the interviews I conducted (30 in total) with notable academics, campaigners and former or serving government officials from each of the five NWS. Several interviewees wished to remain anonymous given their official positions, so I have not named or included a full list of interviewees in this book and, where I use direct quotes, I refer to interviewees alphabetically.[22]

Chapter 8 summarises the strategic relationships between NWS outlined in Chapters 3 to 7, contextualising them in terms of the global nuclear order (including NAS and NNWS), and major socio-political trends and challenges such as climate change. The chapter begins by reviewing possible actions and processes supportive of *institutional democratisation* in the NWS—involving the state, civil society and the public, for example—and how these may be developed to support disarmament action. This involves a discussion of key related issues across NWS with implications for disarmament, such as: recent trends in domestic and international politics, including public opinion, populism and the prospects for a nuclear ban treaty; financial, safety and technological factors involved in nuclear modernisation; and the need to differentiate between states in terms of their responsibility for national nuclear disarmament and a NWFW.

## Notes

1  Rifkind, Malcolm et al. (2013) 'Britain Open to Attack if Trident Is Abandoned', www.telegraph.co.uk, 15 July; Roberts, Brad (2015) *The Case for U.S. Nuclear Weapons in the 21st Century* (Stanford, CA: Stanford University Press); Tertrais, Bruno (2011) *In Defense of Deterrence: The Relevance, Morality and Cost-Effectiveness of Nuclear Weapons* (Paris: IFRI).
2  Brown, Harold and Deutch, John (2007) The Nuclear Disarmament Fantasy, *The Wall Street Journal,* 19 November.
3  Krieger, David and Tutu, Desmond (2014) *We Must End the Madness of Nuclear Weapons,* www.truth-out.org, 20 May; Women's International League for Peace and Freedom (2011) *Disarmament Programme,* http://wilpf.org.
4  Lee, Steven P. (1996) *Morality, Prudence, and Nuclear Weapons* (Cambridge: Cambridge University Press), 319.
5  Datan, Merav, Hill, Felicity, Scheffran, Jürgen, Ware, Alyn (2007) *Securing our Survival: The Case for a Nuclear Weapons Convention* (Cambridge: IPPNW), 139.
6  Schell, Jonathan (2000) *The Fate of the Earth and the Abolition* (Stanford, CA: Stanford University Press), 108.
7  Anthony, Ian (2011) *Irreversibility in Nuclear Disarmament Political, Societal, Legal and Military-Technical Aspects* (Solna: SIPRI), 11.
8  Baum, Seth (2014) Nuclear War, the Black Swan we can Never See, http://thebulletin.org, 21 December.
9  Bulletin of the Atomic Scientists (2020) Closer than Ever: It Is 100 Seconds to Midnight, http://thebulletin.org, January.
10  Weapons of Mass Destruction Commission (2008) *Freeing the World of Nuclear, Biological and Chemical Arms* (Stockholm: Fritzes), 24–25.
11  Kelleher, Catherine McArdle (2011) Introduction, in Kelleher, Catherine McArdle and Reppy, Judith, eds., *Getting to Zero: The Path to Nuclear Disarmament,* (Stanford, CA: Stanford University Press), 7.
12  Wittner, Lawrence (2010) Where Is the Nuclear Abolition Movement Today?, *Disarmament Forum,* Four (Geneva: UNIDIR), 7.
13  Kropotkin, Peter (1939) *Mutual Aid* (London: Penguin); Hofstadter, Richard (1992) *Social Darwinism in American Thought* (Boston, MA: Beacon Press).
14  Russell, Bertrand (1961) *Has Man a Future?* (Harmondsworth: Penguin), 12.
15  Kristensen, Hans M. and Korda, Matt (2020) Status of World Nuclear Forces, https://fas.org, April.
16  Nystuen, Gro and Eide, Stein-Ivar Lothe (2013) *Wanted: Resolute Normative Leadership,* http://nwp.ilpi.org, 5 September; Johnson, Rebecca (2014) Banning Nuclear Weapons: Point of No Return, www.opendemocracy.net, 19 February; Fihn, Beatrice (2015) *A Process Is in the Making,* www.icanw.org, 11 March.
17  Perkovich, George and Acton, James M. eds. (2009) *Abolishing Nuclear Weapons: A Debate* (Washington DC, Carnegie Endowment for International Peace); Primakov, Yevgeny, Moiseyev, Mikhail, Ivanov, Igor and Velikhov, Evgeny (2010) *Start a New Disarmament Plan,* www.in.rbth.com, 22 October.

18  Cimbala, Stephen J. (2009) *Nuclear Weapons and Cooperative Security in the 21st Century: The New Disorder* (London: Routledge), 29.

19  Canberra Commission (1996) *Report on the Elimination of Nuclear Weapons* (Canberra: Commonwealth of Australia); Green, Robert (1999) *Fast Track to Zero Nuclear Weapons* (Cambridge: Middle Powers Initiative); Evans, Gareth and Kawaguchi, Yoriko (2009) *Eliminating Nuclear Threats: A Practical Agenda for Global Policymakers*, http://icnnd.org; Global Zero (2010) *Action Plan*, www.globalzero.org.

20  Stimson Centre (2009) *Unblocking the Road to Zero*, www.stimson.org.

21  Gourevitch, Phillip (1978) The Second Image Reversed: The International Sources of Domestic Politics, *International Organization*, 32(4), Autumn, 881–912; Kapstein, Ethan B. (1995) Is Realism Dead? The Domestic Sources of International Politics, *International Organization*, 49(4), 754.

22  These interviews were made possible by my PhD collaborative studentship, which was provided by the British American Security Information Council (BASIC) and funded by the Economic and Social Research Council (ESRC). Many of these interviews were conducted at the United Nations Headquarters in New York during the 2014 NPT Preparatory Committee.

# Part I

# Understanding nuclear disarmament: theory and practice

# 1 Conceptualising nuclear disarmament

## 1.1 Introduction

How can we best start to understand the challenges NWS face in eliminating their nuclear arsenals? Specifically, how do we identify whom or what is responsible for the NWS's continued possession and modernisation of nuclear weapons, and their refusal to disarm, and how might this situation change? Having reached such understandings, what progressive ideas and action supportive of NWS being able to eliminate their nuclear arsenals should we recommend? Moreover, do such investigations require us to develop a theory of nuclear disarmament, and, if so, what would such a theory look like? In order to answer these and other related questions, it is perhaps best to begin by exploring existing theories of nuclear possession and disarmament, highlighting which areas remain problematic and which require further investigation and explanation. Following this discussion, we may be able to suggest the methods and knowledge necessary to create a more complete theory, incorporating these insights into our research design. Before embarking on this approach, however, there are several issues to take into consideration.

First, we must recognise that because, in the field of nuclear weapons, definitions of both the material and ideational can be hotly contested and are the subject of much political wrangling, we should critically examine how nuclear disarmament has been defined and consider whether it is necessary or sensible to settle on one particular definition. Secondly, existing scholarly theories of nuclear disarmament generally seek to answer either one or more of the following questions:

i) How and why did South Africa and the three former Soviet states (Belarus, Kazakhstan and Ukraine) relinquish their nuclear arsenals?
ii) How and why were Iraq and Libya's nuclear weapons programmes dismantled?
iii) Should, and if so how could and why would, (one or more of) the nine NPS relinquish their nuclear arsenals?

Studies of the first two questions attempt to explain things that have happened and are thus, naturally, based on empirical discussions of evidence. In contrast, studies addressing the third question attempt to explain how something that has not happened and is not happening could take place, and thus deploy both an empirical, and more speculative, approach. Given that my study is of the latter type, it is necessary to consider what evidence I should base the empirical part of my study on and how this evidence should be used to inform the speculative and transformative element. For example, to what extent are studies of NPS nuclear possession, nuclear disarmament by former NAS, or states that had nuclear weapons programmes, useful to considerations of NWS disarmament and how useful are findings from other studies of nuclear weapons decision-making, such as non-proliferation and restraint, that focus on the experience of NNWS?

Thirdly, it should be acknowledged at the outset that this study has a pronounced normative element because I am not just trying to explain NWS decision-making past and present, but wish to contribute to a discussion of how NWS could act in future in order to become FNWS, based on the three-fold assumption that nuclear disarmament is desirable, justifiable and possible. It is thus necessary to consider what the methodological and theoretical implications are of this stance. For example, it is possible to make a heuristic suggestion at this point that my study will require a means of analysing NWS behaviour that can identify the forces enabling the production of nuclear weapons systems and that treats these processes as contingent, dynamic and open to substantial change, even if present conditions make such change seem distant and unlikely. Before beginning to review and assess existing theories of nuclear disarmament it is worth discussing each of these three issues in more depth, both to explore their significance and to try and reach some conclusions about the best way to proceed through this complex terrain.

## 1.2   What is nuclear disarmament?

On the face of it answering the question 'what is nuclear disarmament?' seems straightforward. For example, we may refer to the consensus Final Document of the 2010 NPT Review Conference (RevCon), agreed to by the 189 States Parties to the treaty, which includes an action plan consisting of 'concrete steps for the total elimination of nuclear weapons'.[1] Despite the clear language contained in this and other intergovernmental (for example, United Nations) documents, nuclear disarmament remains a much-contested term, principally due to NWS and NNWS having quite different interpretations of what it—and the NPT itself—means, both from a political and technical point of view.

For example, civil society groups such as Reaching Critical Will point out that progress towards nuclear disarmament must be measured by the 13 Practical Steps agreed by NPT States Parties at the 2000 RevCon for the 'systematic and progressive disarmament of the world's nuclear weapons'.[2] The 2010 NPT RevCon's Final Document reaffirmed the 13 Steps and included a

64-point action plan in order to move forward on achieving the treaty's goals, including a commitment by NWS to reduce the salience—meaning the role and significance—of nuclear weapons in their national security policies.

Importantly, Action 2 of the 2010 RevCon's Final Document states that all members of the NPT 'commit to apply the principles of irreversibility, verifiability and transparency in relation to the implementation of their treaty obligations'.[3] These three principles are designed to increase confidence between states and convince the international community that an NWS has implemented the legal, political and technical measures required for them to eliminate all their nuclear weapons. The aforementioned NPT agreements can also be said to form the strongest existing basis for international discussions on nuclear disarmament. This is because of the number of participating states and the fact that NWS and NNWS reached consensus, lending the NPT process significant legitimacy. Furthermore, as John Burroughs explains, in 1996 the International Court of Justice concluded that 'the threat or use of nuclear weapons is generally illegal' and that there exists an obligation on all states, under Article VI of the NPT, 'to pursue in good faith and bring to a conclusion negotiations leading to nuclear disarmament in all its aspects under strict and effective international control'.[4]

However, as Reaching Critical Will points out, there are, in reality, several problems with the current approach. This is first because the NPT agreements on nuclear disarmament are actually weak, providing 'very few benchmarks to measure progress', and secondly, because 'time lines were removed and the language used is vague and leaves most disputed actions open for interpretation'.[5] Thirdly, efforts to fulfil the requirements of the action plan have so far been 'significantly lacking'.[6] Who or what is responsible for the current impasse? Beatrice Fihn speaks for many NNWS when she argues that responsibility lies with the five 'official' NWS under the NPT: China, France, Russia, the UK and the US. For Fihn, the problem is that NWS 'somehow interpret' Article VI of the NPT as allowing them 'to possess nuclear weapons until they eventually decide to get rid of them'.[7]

The inequality inherent to the NPT was a result of the bipolar global order imposed by the USSR and US at the time the treaty was agreed. The two superpowers decided it was in their shared interest to preserve the status quo by preventing nuclear proliferation through technology denial. Opposing this settlement were the members of the Non-Aligned Movement (NAM), a loose collective of developing nations from the southern hemisphere, and other progressively minded NNWS, which actively pushed for the inclusion of nuclear disarmament obligations in the NPT.

Fihn observes that countries that interpret Article VI as an obligation to negotiate nuclear disarmament are seen by NWS as 'upsetting the strategic balance and even sometimes are blamed for not focusing enough on non-proliferation'.[8] As Mukhatzhanova and Potter point out, however, non-proliferation was 'never a central tenet' of the NAM, since its members wanted to preserve their right to access nuclear technology and focus on abolishing nuclear weapons, as symbols of discrimination and inequality.[9] Based on this analysis, one may conclude that if progress towards the shared goal of nuclear disarmament is to occur, then the attitudes and behaviour of NWS governments will need to significantly change so the agreed steps and action plans can be enacted.

In principle, each of the NWS supports the goal of a NWFW but each has different views on the path—for example, the form and content of multilateral negotiations required to achieve this—based on their strategic aims and objectives. One response by NWS to criticism from civil society and NNWS has therefore been to meet as a group (known as the 'P5') to discuss the implementation of their NPT obligations. More widely, the NWS are each committed to the 'step by step' approach, which consists of negotiations on a series of initial steps towards nuclear disarmament, including further bilateral reductions in nuclear weapons stockpiles between the US and Russia, the agreement of a Fissile Material Cut-Off Treaty (FMCT) and a Comprehensive Test Ban Treaty (CTBT).

Thus, on the one hand, NWS have declared that they are earnestly engaged in discussions, behind closed doors, to create the political conditions that will allow them to realise their Article VI disarmament obligations. Yet, given the high salience nuclear weapons still have in NWS's security doctrines and the significant investments being made to modernise nuclear arsenals, it is clear that nuclear weapons remain 'deeply embedded elements of their strategic calculus', as analysts from the Stockholm Institute Peace Research Institute (SIPRI) note.[10]

Such contemporary disagreements are useful not only to highlight, if it was at all necessary, the controversial nature of nuclear disarmament, but also to begin a discussion of how this study will approach such controversies, given that, as Robert Cox points out, 'there is no theory for itself; theory is always for someone, for some purpose'.[11] The question of whom a theory of nuclear disarmament is for and what its purpose is may seem straightforward. Efforts to ban the bomb have, historically, consisted of popular struggles led by people of all nations, with NAM members featuring prominently, and may be said to comprise, after Wittner, the 'world nuclear disarmament movement'.[12]

The historic purpose of what I shall hereafter refer to as the *disarmament first* approach to the abolition debate is the elimination of nuclear arsenals for principally moral reasons. Whilst abolitionists may disagree on several questions, for example, which strategy should be adopted to build momentum for nuclear disarmament, they commonly argue that any use of nuclear weapons would result in indiscriminate suffering and destruction of life on a huge scale—as experienced in the US atomic bombings of Japan in 1945.[13] Abolitionists also draw on a variety of other evidence, using humanitarian, legal, security, environmental and economic arguments, as well as supportive national and global public opinion, in order to explain why nuclear disarmament is justifiable and realisable.

In general, there are several areas of agreement within *disarmament first* regarding the purpose and requirements of nuclear disarmament. The first of these, as described above, is the moral and humanitarian justification for abolishing nuclear weapons as the necessary precursor to the establishment of common security amongst nations.[14] Secondly, given their moral rejection of nuclear weapons, proponents of this approach advocate a range of unilateral, bilateral and multilateral nuclear disarmament measures, placing the responsibility for such urgent action squarely on the shoulders of each NPS. Thirdly, nuclear disarmament is seen as needing to be irreversible and permanent. Today, campaigners from several groups and coalitions thus engage in activism to pressure governments to support a Nuclear Weapons Convention or the Treaty on the Prohibition of Nuclear Weapons (TPNW). Such efforts led to the 50th ratification of the TPNW being deposited in October 2020, meaning that the treaty entered into force on 22nd January 2021.

It is important to highlight the *disarmament first* approach at the outset of this study since a scholarly investigation of the political conditions necessary for NWS nuclear disarmament cannot ignore the history that preceded it and which it is a part of. Such a study should therefore aim to produce knowledge that will be of use and accessible to anyone interested in NWS nuclear disarmament and the realisation of a NWFW. Yet this issue is complicated by the fact that, in recent years, prominent figures from across the political spectrum in NWS have voiced support for the idea of a NWFW. Most notable is the apparent change of heart regarding nuclear weapons by several former statesmen and women (mainly representing British and US elites) who could be described as Cold Warriors, hawks or political realists. In particular, the phenomenon whereby figures from the US political establishment such as George Schulz, William Perry, Henry Kissinger, Sam Nunn—and many others—have come out in favour of 'a world without nuclear weapons' requires close critical examination.[15]

Recent elite contributions to the debate regarding the possibility of a NWFW are largely based on a recognition that nuclear weapons—once exclusively the weapons of the strong—have become potential weapons of the weak, threatening the great powers and changing international power dynamics. Indeed, David Cortright and Raimo Väyrynen have observed that the proliferation of nuclear weapons 'is both a cause and consequence of the growing decentralisation and multipolarity of international relations'. The 'effectiveness of deterrence and the old bipolar international order centred on Russia and the United States' is thus crumbling, so that where there was once stability there is now uncertainty.[16]

Similarly, Alistair Young points out that, whilst the US has preponderant power 'across the range of key power resources—military, economic and technological', since the 'latter part of the 2000s' the distribution of power appears to be shifting, following the US's disastrous occupations of Iraq and Afghanistan, the 2008 financial crisis and the 'increased economic importance and greater assertiveness of China, Russia, India and Brazil'.[17] Nuclear proliferation is also feared because it increases the risk that non-state actors will acquire nuclear weapons for terrorist purposes. Recent calls for action on nuclear disarmament from within the US and other NWS elites have therefore, in large part, been driven by the realisation that they must be seen to be acting to reduce nuclear dangers—and help prevent the possibility of nuclear weapons being used against their nations.

It is worth noting at this point that Russian and US governments have, since nuclear weapons were first produced, discussed plans for abolition. These include the unsuccessful Acheson-Lilienthal report and the subsequent Baruch plan, proposed by the US, and the Soviet Gromyko plan—all from the beginning of the Cold War—and, more recently, as the Cold War was ending, the Reykjavik meetings between the US and USSR that led to massive reductions in both sides' nuclear arsenals. Yet Raymond Garthoff argues that Reykjavik can be seen as a 'spectacular missed opportunity' for the final realisation of nuclear disarmament.[18] Similarly, former US General George Lee Butler has described the period following the end of the Cold War as a 'priceless opportunity' for the US and Russia to deal with the problems posed by nuclear weapons but one which 'got stepped all over'.[19]

Clearly there is much one might learn from these historical episodes, both to avoid repeating past mistakes and better understand the present. For example, what are the differences between advocacy for nuclear disarmament stemming from grassroots social movements on the one hand and elite groups on the other? What are these group's respective strengths and weaknesses, how are they distributed across NWS and beyond and what have been their successes and failures regarding nuclear disarmament? Moreover, how can we characterise these group's past relationships and can, and if so how, might they work together to build political momentum for action on nuclear disarmament now and in the future?

Whilst recognising the importance of such questions, I will leave them to be addressed later in this study. This is principally because, however we view the past and its successes and failures, we must first identify which contemporary problems require attention. The principal issue for this study is that, whilst opportunities for nuclear disarmament remain open, momentum is currently in the opposite direction given the continual modernisation and reproduction of nuclear weapons by all NWS. Recognising the urgency of this situation and the need to provide intellectual weight to the debate, several scholars, including George Perkovich and James Acton, and Cortright and Väyrynen, have produced detailed works discussing the interlinking political conditions, legal instruments and technical requirements of nuclear disarmament and a NWFW.[20]

To revisit Cox's point, it is pertinent to consider whom these and other proposals and theories are for. Are they, for example, addressing a technocratic, military or political elite or are efforts being taken to reach a global, public audience? In order to properly answer questions about the practical and political significance of such studies, it is necessary to provide some context and discuss the wider theoretical aspects of nuclear disarmament. To begin doing so I return now to the question 'what is nuclear disarmament?' simply because, if we do not have a clear grasp of the term and its varied usage and meanings, it will not be possible to understand the changes in attitudes and behaviour potentially required from decision-makers and institutions in and between NWS.

### Technical, legal and political aspects of nuclear disarmament

In order to enrich our understanding of this issue it is necessary to look beyond the official documents of the NPT and United Nations and delve into the wide and varied literature concerning the enigma of nuclear disarmament. This literature serves either one or both of the following functions: i) to review the history of efforts to eliminate nuclear weapons from 1945 to the present, ii) to speculate on possible future efforts to eliminate nuclear weapons. Given this inquiry's focus on the five NWS and limits of time and space, I principally concentrate on comparing and contrasting works that fulfil the latter function. As will become apparent, since the mid-2000s there has been a notable upsurge in such publications, the causes and consequences of which shall be considered. In any case, the history of nuclear disarmament will not be neglected as many of the recent analyses examine the historical record, so this will form a significant part of our discussion.

Nuclear disarmament as a term has been used to refer to several different 'levels of goals' including: i) freezing, ii) reducing and iii) eliminating some or all nuclear weapons, as well as iv) the creation of a NWFW.[21] The first three of these goals can be conducted by one state or amongst several different collections of states, whereas a NWFW would apply universally. For example, unilateral disarmament may involve one state eliminating its nuclear weapons without seeking equivalent concessions from its actual or potential rivals. Beyond unilateral initiatives, disarmament negotiations resulting in participating states eliminating their nuclear arsenals could take place—separately or together—at a bilateral (involving two NWS/NAS) or multilateral level (involving several NWS/NAS).

These technical points are of importance for my study given that I am investigating the contribution the five NWS could make to a NWFW by completing the elimination of their nuclear weapons. In this sense nuclear disarmament, as a term, is adaptive and provides flexibility when discussing this topic as it can be contracted or expanded as required, for example, to include NAS and the different phases of elimination, as will be necessary for the transition to a NWFW. Notably, a Chatham House study published in 2000 entitled *Nuclear Weapons Policy at the Crossroads* argued that 'little detailed analysis has been undertaken to determine precisely what a world without nuclear weapons would involve' and that a 'consensus' on how a NWFW could be achieved was necessary if NWS were to move 'below a minimal deterrent'.[22] It is pleasing to note that since then a variety of studies, some of which are discussed below, have been produced that begin to develop the knowledge, including the technical understanding and legal instruments, required by global nuclear disarmament.

What is principally lacking therefore, as has been widely recognised, is the sustained political determination in and amongst NWS to drive the project of nuclear disarmament forward, both domestically and through international cooperation. The contemporary political challenges of nuclear disarmament thus form my principal focus and begin to be explored in Chapter 2. Yet, in order not to neglect the importance of technical perspectives on this question, and to be mindful of how expert practitioners have defined nuclear disarmament given the political implications of such definitions, it is worth now briefly outlining some of the work that groups, including VERTIC—a leading verification and monitoring NGO—have done on this issue.

In their report *Irreversibility in Nuclear Disarmament: Practical Steps Against Nuclear Rearmament,* VERTIC researchers define nuclear disarmament as a state in which 'the process of disarming has been fully completed and no nuclear weapons remain' and consider how this state can be 'locked-in'.[23] The early stages of disarmament will need to develop verification systems covering warheads, delivery systems and fissile materials.[24] There are potential scales of irreversibility here so that a considerably 'higher' level of disarmament would involve measures directed toward both a state's warhead stockpile and its supporting nuclear warhead production complex.[25]

Overall the question of irreversibility is important because, as Scott Sagan argues, in a NWFW former possessor states would be 'more latent' than states which 'did not have their technological expertise or operational experience'. Owing to the fact that the five NWS will always be NWS according to the NPT, it therefore makes sense to use Sagan's recommended designation of Former Nuclear Weapon State (FNWS) for NWS that in future fully complete the nuclear disarmament process. Secondly, rather than focusing on scrapping a particular weapons system, FNWS status—as an objective—conveys more appropriately the wider political implications for NWS if they are to live up to their international responsibilities and disarm irreversibly, verifiably and transparently.[26]

Every level up the irreversibility scale makes rearmament more difficult, requiring more resources and time—equally, more money, time and equipment is required to attain a higher disarmament level. For Justin Alger and Trevor Findlay, there is a lack of precise information concerning the costs of disarmament—a 'void' which, they claim, needs to be filled to 'advance the discussion about nuclear disarmament beyond the philosophical to the practical'. Despite this gap, it is clear to them that such costs will be spread over several decades and be incurred at different points, so that:

> dismantlement and disposition costs will come in the early stages, along with strengthening of nuclear safeguards, while verification costs will

ramp up as the process nears zero and becomes politically and strategically more sensitive.[27]

As Findlay points out, the verification and compliance regime for a NWFW 'will need to be more effective than any disarmament arrangement hitherto envisaged'.[28] This is necessary to cope with fears of breakout, which is when a state is suddenly revealed to have a previously hidden nuclear arsenal, produced new weapons or sufficient weapons-usable fissile material (highly enriched uranium or plutonium). In addition, the authors of the report *Unmaking the Bomb: A Fissile Material Approach to Nuclear Disarmament and Nonproliferation* argue that, because a fissile material free world would make a NWFW more stable, it is necessary for the international community to engage in a step-by-step process to:

> cap, reduce, and eventually eliminate the global stockpile of about 1,900 tons of weapon-usable fissile material including material in weapons or recovered from dismantled weapons, the plutonium used in civilian nuclear power programs, and the HEU in military and civilian research and naval reactor fuel stockpiles.[29]

Findlay notes that, while meeting the technical requirements for nuclear disarmament and a NWFW is a 'tall order', practical experience of disarmament and improving technologies mean that it is possible. He also argues that the 'good relationships' between states that will facilitate the negotiation of a nuclear disarmament treaty will also permit the construction of an appropriate verification and compliance system.[30] Whilst it is generally acknowledged that the creation of a verification regime able to support a NWFW is, for political and technical reasons, an immense task, examples exist that can be built on. As researchers at SIPRI note, such ideas are needed to realise disarmament action in the short term and lay the foundations for more far-reaching measures, such as an operational TPNW.[31]

Examples of successful verification include the International Atomic Energy Agency's verification of the dismantlement of South Africa's nuclear arsenal, the experience of arms control treaties between the US and Russia, as well as recent technical collaboration on warhead dismantlement verification between the UK and Norway. In addition, the UK and US began a project in 2000 on technical cooperation concerning the multilateral verification of warhead dismantlement, with a focus on how to balance 'information protection and information sufficiency' in an 'effective monitoring and verification regime'.[32] Other projects of this type include the International Partnership on Nuclear Disarmament Verification and the Quad Nuclear Verification Partnership, involving the UK, Norway, Sweden and the US. The importance of these recent efforts is in their focus on technical and institutional issues concerning warhead elimination from a multilateral perspective. Wyn Bowen and his co-authors note that previous verification initiatives have concentrated on delivery vehicles, with the elimination of warheads 'strictly controlled and overseen by national authorities'.[33]

Pavel Podvig, Ryan Snyder and Wilfred Wan observe that the key verification challenges to overcome for disarmament concern the 'sensitive nature of nuclear weapons', so that data on 'weapon design and the composition of fissile materials' must be protected. They therefore propose an 'arrangement' focusing on 'verifying the absence of nuclear weapons and infrastructure for their deployment', which could appropriately be applied to cover 'a certain territory' or 'class of weapon systems'. The focus on proving absence, the authors argue, would be beneficial for several reasons, including the avoidance of several of the 'most complex and daunting problems associated with nuclear disarmament verification' and the facilitation of the use of 'tools and techniques otherwise unavailable when the presence of nuclear weapons is a possibility'. The successful deployment of this arrangement would, their study concludes, help demonstrate the feasibility of 'practical disarmament measures' and build confidence that more 'comprehensive' steps can be taken in the long term.[34]

Elsewhere, the Canberra Commission's report on eliminating nuclear weapons argues that disarmament should be approached 'as a series of phased verified reductions that allow states to satisfy themselves, at each stage of the process, that further movement toward elimination can be made safely and securely'.[35] Michael Mazarr and other scholars, in a publication complementing the findings of the Canberra Commission report, propose that 'removing all nuclear weapons from operational status and placing them in a dismantled "virtual" condition' would be an important initial phase, prior to disarmament, in 'pushing nuclear weapons to the margins of world politics'.[36] Quantitative and qualitative reductions, covering the numbers *and* salience of nuclear weapons, are thus an important joint consideration for the NWS, particularly in terms of how they can meet their NPT disarmament obligations.

Aspects of the NWS's current nuclear policies which could be changed to reduce the salience of nuclear weapons (also potentially as moves towards a virtual nuclear arsenal as part of the transition to disarmament) concern: i) acquisition: meaning what kit is bought and owned—a particularly important issue when each NWS is pursuing nuclear modernisation; ii) declaratory: public statements about the role of nuclear weapons; iii) deployment: how nuclear weapons are arranged and positioned; iv) employment: the circumstances and ways in which a government plans to use nuclear weapons to achieve its strategic aims.

An examination of how each of the five NWS might engage in such steps is presented in Chapters 3 to 7, suffice to say here that disarmament can also be seen as a learning process, involving far greater transparency regarding military capabilities and intentions, so that states gradually 'abandon secrecy', reduce uncertainty and continually improve their appreciation and knowledge of the legal and technical requirements of getting to zero.[37] This brief foray into the technical aspects of nuclear disarmament shows us that, given the implications of these decisions for national sovereignty—not least in terms of how highly sensitive information should be handled—political agreements with wide international support will need to be reached on the appropriate scale of irreversibility, and the accompanying transparency, verification and compliance regime, for each of the NWS if they are to satisfactorily eliminate their nuclear arsenals.

Furthermore, if, as Findlay argues, amicable international relations are a necessary condition both of the process leading to disarmament and the maintenance of a NWFW, we should investigate the form these relationships might need to take. As I discuss in Chapter 2, several authors have addressed this question, some considering how cooperation and trust may be developed between NWS, whilst others emphasise the need for some form of political agreement, union or concert between the great powers, alongside disarmament action. In addition, NWS governments will need to be persuaded that the benefits of disarming outweigh the costs and ensure their citizenries are well informed about these matters.

## 1.3 Nuclear weapons decision-making: past, present and future

The purpose of the theory under consideration here is to contribute to an understanding of the challenges involved in the five NWS achieving nuclear disarmament. The question then is, how wide should we cast our net in order to develop such a theory? By this I mean, what data do we need and what case studies should be used? This is a methodological problem because the literature on nuclear weapons is vast, covering several decades and a wide range of actors. In order to more easily navigate through the literature and identify what areas need to be covered and in what depth, we may split the literature into two sections.

First, as identified above, there is a large selection of historical studies of nuclear weapons, which mainly consider what has happened. These theories can differ significantly in their methods and conclusions but share the good fortune of being able to build concepts from quantitative and qualitative data regarding the past and present behaviour, ideas and preferences of a range of actors. The four main categories within this first section includes theories of: i) nuclear possession: how and why NPS acquire, manage and use nuclear weapons as they do; ii) nuclear restraint: how and why NNWS choose not to acquire nuclear weapons when they have the opportunity and incentive to do so; iii) nuclear arms control: how and why NPS and NNWS place restrictions on the development, production, stockpiling, use and proliferation of nuclear weapons; iv) nuclear disarmament: how and why former NAS and states with nuclear weapons programmes eliminated their nuclear weapons or were disarmed.

Secondly, there is a much smaller selection of studies discussing the challenges of nuclear disarmament and/or a NWFW, including whether this is a desirable and realisable goal. These tend to focus on particular NPS, pairings or groups of NPS or NPS as a whole, considering the problems posed by nuclear disarmament in the context of the wider international political environment. Such studies are based on empirical research but are, albeit to a different degree, unavoidably speculative. They are empirical because they tend to draw on evidence and theories both regarding nuclear possession and arms control directly involving NPS, in addition to nuclear possession, restraint, arms control and disarmament involving NNWS. They are speculative because they propose future courses of action by NPS, for example, behaviour that is/is not conducive to nuclear disarmament.

Such conjecture concerning nuclear disarmament is problematic because, as Harald Müller points out, the evolving interaction between several different actors over time cannot be reliably predicted: 'as conditions change, so do the structures of opportunity. New options, unthinkable at the beginning, become a serious possibility.'[38] It is thus vital to develop ideas and proposals that are rooted in evidence and experience and which, as Müller suggests, are flexible and able to adjust to changing realities, in order to look 'far ahead' but not spoil the process by 'fixing strategies that should be subject to continuous adaptation because of changing circumstances'.[39]

In order to draw on the widest empirical base to inform justifiable speculation on the political conditions necessary for NWS nuclear disarmament and the means by which it may then be realised, this study would ideally review studies representing each of the four main categories in the first section in order to get a rounded sense of the subject. However, because this study faced limits of time and resources, it was also necessary to impose boundaries on such a review. I therefore use discussions concerning: i) nuclear possession by NAS, ii) restraint, iii) arms control, iv) nuclear disarmament—particularly how and why NAS have eliminated their nuclear weapons—as secondary material to highlight conceptual differences as and when necessary in order to principally focus in this study on works discussing NWS nuclear possession and development. The case for a consideration of the literature on Ukrainian and South African disarmament, and the establishment of nuclear weapon free zones, was perhaps the strongest. Ultimately however, I decided that the unique nature of these cases and the fact that NWS disarmament will be on a much greater scale limited the extent to which political conclusions can be drawn from these cases of relevance to NWS, so justifying their omission.

With regard to the second section of the nuclear weapons literature, concerning the contemporary challenges of nuclear disarmament and/or a NWFW, this is a question of selecting from an expanding, but still small collection of works, some of which are of much greater depth and significance than others. This section of the literature principally considers historical case studies in order to inform contemporary and future challenges associated with nuclear disarmament. I begin my substantive discussion of both sections of the nuclear weapons literature in Chapter 2, primarily in order to begin assessing the explanatory power of existing theories of nuclear possession and disarmament. Before embarking on this task, it is useful to provide a brief consideration of how the nuclear disarmament debate relates to a third and highly important section of the relevant literature—namely, political theory.

## 1.4 How does nuclear disarmament relate to political theory?

As discussed above, nuclear disarmament exists in many works as a practical issue that raises a number of political, legal and technical challenges requiring appropriate solutions. These works are largely produced by authors attached to different types of institutions—both governmental and non-governmental—including, in the latter case, campaign groups, research bodies and academia. Yet nuclear disarmament is also the subject of much controversy within another field, that of political theory. Discussions of political theory almost exclusively take place within academia, and generally debate nuclear issues in terms of how they relate to the dominant schools of thought, or what Mearsheimer and Walt refer to as the 'grand theories' of international relations.[40]

As these authors note, 'Grand theories such as realism or liberalism purport to explain broad patterns of state behavior, while so-called middle-range theories focus on more narrowly defined phenomena like economic sanctions, coercion, and deterrence.'[41] Following this categorisation, the theory of nuclear disarmament I develop in this study is a middle-range, rather than a grand theory. Moreover, as we shall see in Chapter 2, nuclear issues—including disarmament—are mainly discussed in the scholarly literature in relation to explanations of state behaviour at the international level, based on the assumptions of grand theory, particularly realism. The theory developed in this study is thus, principally, a response to and critique of how realism explains and understands nuclear possession and disarmament.

Before outlining the meaning and significance of realist and other international relations theory regarding these matters, it is necessary to explain why this study will discuss nuclear disarmament in relation to political theory at all. First, reviewing the theoretical literature allows us to identify what claims and approaches already exist regarding the causes and consequences of nuclear disarmament, and assess their strengths and weaknesses. This is also an important task, given that there are strong criticisms of and objections to nuclear disarmament, on moral, political and practical grounds, within the theoretical literature. Given the social and political influence, and weight of these objections, it is important for a study of this type—based on a normative interest in nuclear disarmament—to provide a coherent and well-evidenced critique of works that support nuclear possession and question the legitimacy of nuclear disarmament. Furthermore, engaging with ideas from across the political and theoretical spectrum should enable us to craft a more coherent and robust theory of NWS nuclear disarmament and, more widely, contribute to political and international relations theory by highlighting any gaps and problems within it.

In order to assess whether existing theories are able to explain the current disarmament impasse and propose effective action supportive of disarmament, Chapter 2 therefore conducts an in-depth discussion of realist thought and scholarly critiques of realism. Having identified the strengths and weaknesses of existing theory concerning the causes and consequences of nuclear possession and disarmament, I then propose alternative ways of analysing nuclear politics as a basis for pinpointing what political change is necessary if NWS nuclear disarmament is to be realised.

## Notes

1  United Nations (2010) *Final Document: Review Conference of the Parties to the Treaty on the Non-Proliferation of Nuclear Weapons,* www.un.org, 19.
2  Reaching Critical Will (2020) *Nuclear Non-Proliferation Treaty*, www.reachingcriticalwill.org.
3  United Nations, *Final Document*, 20.
4  Burroughs, John (1998) *The Legality of Threat or Use of Nuclear Weapons: A Guide to the Historic Opinion of the International Court of Justice* (Berlin: Lit Verlag), 48–51.
5  Fihn, Beatrice (2013) Multilateral Treaty-Based Commitments and Obligations, in *NGO Presentations to the Open-Ended Working Group on Taking Forward Multilateral Nuclear Disarmament Negotiations* (New York: United Nations), 4.
6  Reaching Critical Will (2020) *Nuclear Non-Proliferation Treaty,* www.reachingcriticalwill.org.
7  Fihn, Multilateral Treaty-Based Commitments, 5.
8  Ibid.
9  Mukhatzhanova, Gaukhar and Potter, William (2011) Nuclear Politics and the Non-Aligned Movement, *Adelphi Series*, 51(427), 40.
10  SIPRI (2014) SIPRI Launches World Nuclear Forces Data, www.sipri.org, 16 June.
11  Cox, Robert (2010) Robert Cox on World Orders, Historical Change, and the Purpose of Theory in International Relations, www.theory-talks.org, 12 March.
12  Wittner, Lawrence (2009) *Confronting the Bomb: A Short History of the World Nuclear Disarmament Movement* (Redwood City, CA: Stanford University Press).
13  Campaign for Nuclear Disarmament (2012) *The Bombing of Hiroshima and Nagasaki*, www.cnduk.org.
14  Ibid.
15  Nuclear Security Project (2011) *Vision and Steps*, www.nuclearsecurityproject.org.
16  Cortright, David and Väyrynen, Raimo (2009) *Towards Nuclear Zero* (London: Routledge), 15–16.
17  Young, Alistair (2010) Perspectives on the Changing Global Distribution of Power: Concepts and Context, *Politics*, 30(S1), 3.
18  Garthoff, Raymond L. (1994) *The Great Transition: American–Soviet Relations and the End of the Cold War* (Washington DC: Brookings), 285.
19  Schell, Jonathan (1998) *The Gift of Time* (London, Granta), 207.
20  Perkovich, George and Acton, James M., eds. (2009) *Abolishing Nuclear Weapons: A Debate* (Washington, DC, Carnegie Endowment for International Peace); Cortright and Väyrynen, *Towards Nuclear Zero* (London: Routledge).

21 Galtung, Johann (1984) *There Are Alternatives! Four Roads to Peace and Security* (Nottingham: Spokesman), 125.

22 Howlett, Daryl, Ogilvie-White, Tanya, Simpson, John and Taylor, Emily (2000) *Nuclear Weapons Policy at the Crossroads* (London: Chatham House), 47–9.

23 Cliff, David, Elbathimy, Hassan and Persbo, Andreas (2011) *Irreversibility in Nuclear Disarmament: Practical Steps Against Nuclear Rearmament* (London: VERTIC), 6.

24 Howlett et al., *Nuclear Weapons Policy*, 52.

25 Cliff, *Irreversibility*, 16.

26 Sagan, Scott (2010) Shared Responsibilities for Nuclear Disarmament, in Sagan, Scott, *Shared Responsibilities for Nuclear Disarmament: A Global Debate* (Cambridge: American Academy of Arts and Sciences), 5, 12.

27 Alger, Justin and Findlay, Trevor (2009) *The Costs of Nuclear Disarmament*, http://carleton.ca, September, 3, 21.

28 Findlay, Trevor (2003) *Verification of a Nuclear Weapon-Free World* (London: VERTIC), 2.

29 Feiveson, Harold A, Glaser, Alexander, Mian, Zia von Hippel, Frank N. (2014) *Unmaking the Bomb: A Fissile Material Approach to Nuclear Disarmament and Nonproliferation* (Boston, MA: MIT), 2.

30 Findlay, *Verification*, 2.

31 Erästö, Tytti, Komžaitė, Ugnė and Topychkanov, Petr (2019) *Operationalizing Nuclear Disarmament Verification*, www.sipri.org, April, 1.

32 UK Government (2010) *The United Kingdom–Norway Initiative: Research into the Verification of Nuclear Warhead Dismantlement*, www.gov.uk; Feiveson et al., *Unmaking the Bomb*, 2; National Nuclear Security Administration (2015) *Joint U.S.-U.K. Report on Technical Cooperation for Arms Control*, www.nnsa.energy. gov, 2.

33 Bowen, Wyn Q., Elbahtimy, Hassan, Hobbs, Christopher and Moran, Matthew (2018) *Trust in Nuclear Disarmament Verification* (London: Palgrave Macmillan), 56.

34 Podvig, Pavel, Snyder, Ryan and Wan, Wilfred (2018) *Evidence of Absence: Verifying the Removal of Nuclear Weapons* (Geneva: UNIDIR), 1–3.

35 Canberra Commission (1996) *Report on the Elimination of Nuclear Weapons* (Canberra: Commonwealth of Australia), 10.

36 Mazarr, Michael J. (1997) *Nuclear Weapons in a Transformed World: The Challenge of Virtual Nuclear Arsenals* (New York: St Martin's Press), 4.

37 Schaper, Annette and Müller, Harald (2008) Torn Apart: Nuclear Secrecy and Openness in Democratic Nuclear-Weapon States, in Evangelista, Matthew, Muller, Harald and Schornig, Niklas, eds., *Democracy and Security: Preferences, Norms and Policy-Making* (Abingdon: Routledge), 155.

38 Müller, Harald (2009) The Importance of Framework Conditions, in Perkovich and Acton, *Abolishing Nuclear Weapons*, 174.

39 Ibid., 177.

40 Mearsheimer, John J. and Walt, Stephen M. (2012) *Leaving Theory Behind: Why Hypothesis Testing has Become Bad for IR* (Cambridge, MA: Harvard Kennedy School, Faculty Research Working Paper Series, 7 December), 3.

41 Ibid., 10.

# 2 Assessing theories of nuclear possession and disarmament

## 2.1 Introduction

This chapter will discuss existing approaches to nuclear possession and disarmament in realist thought and scholarly critiques of realism, with three main purposes in mind. The first is to assess the extent to which such works provide useful insights into the causes of nuclear possession as well as the goals, processes and indicators of and conditions for nuclear disarmament, which may be applied to the problem of how to eliminate the NWS's nuclear weapons. The second is to address what several scholars see as the major issue of how international anarchy drives conflict and stifles cooperation between states. According to this perspective, because nuclear disarmament requires significant international cooperation, the substantial obstacles to this posed by anarchy—such as fear and uncertainty—need to be addressed. The third is, where necessary and useful, to improve on existing approaches and positions by developing a new theoretical approach that can better explain and respond to the challenges of achieving NWS nuclear disarmament.

The former two areas will be addressed in the following way. First, this chapter reviews key relevant works from the realist tradition and scholarly critiques of realism that elucidate on the claim that the invention of nuclear weapons heralded a seismic shift in the nature of warfare and international relations—a 'nuclear revolution'. For Robert Jervis, this 'revolution' refers to the fact that, no matter how powerful a nation's military is, nuclear weapons are impossible to defend against. Therefore, as he argues (following Thomas Schelling's view), what is significant about nuclear weapons is the concept of 'mutual kill', whereby 'the side that is "losing" by various measures of military capability can inflict unprecedented destruction on the side that is "winning" as easily as the "winner" can do this to the "loser"' Given the extreme and singular power of these weapons, 'brute force', for Jervis, has thus been replaced 'by coercion, or, as it is more frequently put, of defense by deterrence'.[1]

To properly review the meanings ascribed to the advent of nuclear weapons, a range of international relations thought is discussed, including, in non-chronological order: i) Offensive Realism in the work of John Mearsheimer, ii) Defensive Realism in the work of Robert Jervis and Charles Glaser, iii)

Structural Realism in the work of Kenneth Waltz, iv) Institutional theory and International Cooperation in the work of Robert Keohane and Nicholas Wheeler respectively, v) World Government and Republican Security Theory in the work of Campbell Craig and Daniel Deudney respectively.

I discuss these different approaches, some of which are placed together within a section where appropriate, by first reviewing their general perspective on international relations, focusing on their understanding of how the anarchic international state system functions and relating this to the meaning of the nuclear revolution. For example, the former three realist groups in particular could be said to broadly fit within what Scott Sagan terms the 'Security Model', which focuses on how 'international threats' to state's 'sovereignty and national security', specifically, 'rival states' acquiring nuclear weapons, drive decision-makers in 'strong states' to seek the bomb.[2] For some authors—such as Mearsheimer and Waltz—the demands of anarchy thus naturally lead great powers to see nuclear possession as essential for their security and survival. According to this logic, nuclear disarmament is undesirable and unrealisable, both because it threatens existing peace and stability and necessitates unachievable degrees of international cooperation.

Other authors I review, such as Craig and Deudney, observe that the nuclear revolution poses a deep challenge to the traditional tenets of realist thought because NWS rely on nuclear weapons for security yet these weapons threaten the existence of these and all other states given the extreme dangers they pose. According to this position, the nuclear revolution therefore requires international relations theory to move beyond realism and accept that nation states need to be transcended by some form of global political organisation. However, these authors also recognise that replacing global anarchy with global hierarchy risks introducing tyranny, so that some other solution must be sought. The work of Keohane and Wheeler is relevant here as they propose that states may avoid the conflict so prevalent under anarchy through adopting cooperative measures. Keohane focuses on the development of international institutions as a way of motivating states to work together to solve common challenges, including those with a military or security focus, whilst Wheeler examines how states may develop trust, to facilitate—amongst other things—progress on nuclear abolition.

In addition to focusing on the international state system, I pay particular attention to how each author under discussion treats the US case, including in relation to nuclear issues. This is necessary given the US's singular power, the fact that nearly all of the main authors reviewed are US-based, and because each devotes significant space in their analysis to the US's strategic behaviour. I then review how each of these authors have viewed the project of nuclear disarmament, for example, whether they see it as a desirable and realistic enterprise, the political and security problems they see it posing and any ideas or proposals they discuss which could be supportive of it. In addition, at appropriate points in the analysis, I draw on the thought of other key international relations thinkers, such as John Herz and Thomas Schelling, to explore particularly relevant ideas and debates.

Having reviewed these different approaches to nuclear matters, I identify several significant problems in the thinking of the constituent authors on these issues, including disarmament. Some of these problems, which are more common and more pronounced than others for each author, include:

- The adoption of a nationalistic approach that prioritises US security interests above all others, leading to support for continued nuclear possession by the US in some shape or form.
- A normative bias in favour of nuclear possession that precludes a substantive consideration of the need for and benefits of nuclear disarmament, including a failure to properly consider the substantive legal, moral and security arguments for disarmament and the popular support for disarmament action.
- The absence of a political analysis that considers the domestic impact of nuclear possession, and which relates domestic politics to a state's international behaviour on nuclear issues.

Such absences and deficiencies raise several analytical gaps that have to be filled in order to develop a compelling theory of nuclear disarmament, including:

- The need for detailed empirical data concerning nuclear possession in and between the NWS to support an analysis of why these states persist in their possession of the bomb that addresses domestic political dynamics.
- The need to identify a corresponding theory of change for nuclear disarmament that is legitimate and based on principles of democracy, transparency and accountability and that also specifies the actors and processes that will be able to realise such change.
- The need for a more objective approach to the politics of nuclear possession and disarmament, for example, one based on justice and universalism that can appropriately identify and differentiate between each NWS's responsibility for nuclear disarmament.

Having identified the key gaps in and problems with the mainstream literature, I propose that, in order to improve on it, we need to develop a new middle-range theory, which I term *institutional democratisation*. This approach, which is based on a normative stance supportive of nuclear disarmament, builds on insights from global governance and civil society literatures as well as social and political theory, including Scott Sagan's domestic politics model, which focuses on the domestic drivers of nuclear acquisition. *Institutional democratisation* can be summarised as the idea that the democratic deficit in the five NWS—including, crucially, the US given its singular influence, power and global reach—needs to be addressed if NWS are to completely eliminate their nuclear arsenals.

This approach is necessary because nuclear weapons systems can be seen as a shared scientific and technological culture amongst nuclear elites in the global north, with highly centralised decision-making structures for nuclear threats and detonation. The secrecy surrounding the policies guiding these systems is also incompatible with and deeply corrosive to the spirit and functioning of democratic institutions, where they exist. Nuclear disarmament therefore requires, to varying degrees, popular social movements to democratise the state in each NWS. This could mean different things in practice according to the domestic circumstances of each NWS, but centres around the need to reform state institutions so that defence and foreign policy generally, and nuclear weapons decision-making in particular, is under civilian and democratic control. Questions, such as whether democratisation would automatically lead to disarmament, will also be important to consider here, as will the types of civil society and other popular activism that would best support disarmament.

Having developed the concept of *institutional democratisation* in this chapter, in Part II I identify the strengths and weaknesses of this theory's ability to explain nuclear politics by conducting a historical investigation of key debates and policies concerning nuclear possession and disarmament in and between the NWS. As well as providing some means of testing *institutional democratisation* as a theory, evidence and ideas are gathered on the necessary indicators and conditions for nuclear disarmament. This process also identifies and responds to the different political setups of each NWS so that each of their unique nuclear and political histories and regimes are taken into account.

It is also important to state at the outset that in carrying out this investigation, I directly engage with the methodological problem of how to judge whether one theory may perform better than another given the speculative nature of nuclear disarmament—as something that has not happened and is not happening. When it comes to the conditions, goals, processes and indicators of and for nuclear disarmament, the main problem we therefore have is that our ideas and proposals are unavoidably incomplete and speculative. Any demand that existing positions relating to the question of nuclear disarmament be analysed and then alternative models or theories be developed as rival positions to show how one performs better than the other, may thus be useful, but only to a limited extent, for the reasons outlined above.

## 2.2 Offensive realism: John Mearsheimer

In an interview with *International Relations*, John Mearsheimer described his self-confessedly 'pessimistic' approach to politics, stating that 'I am an offensive realist who believes that war is a legitimate instrument of statecraft and that states should maximize their relative power.'[3] Mearsheimer developed his 'descriptive' and 'prescriptive' theory of offensive realism in his major work, *The Tragedy of Great Power Politics,* in which it is argued that international conflict has been so prevalent because great powers are power maximisers, each sharing 'hegemony as their final goal'.[4] This, for Mearsheimer, is how he believes such states not only mostly do but also should act in order to survive and thrive. Status quo powers thus only exist when a hegemon 'wants to maintain its dominating position over potential rivals'.[5] He explains this power competition by outlining five 'bedrock assumptions' about the international system. First, following other structural realist thought, Mearsheimer sees this system as 'anarchic', which stems from the lack of a 'central organizing power'. Secondly, 'great powers inherently possess some offensive military capability'. Thirdly, 'states can never be certain about other states' intentions'. Fourthly, 'survival is the primary goal of great powers'. Fifthly, 'great powers are rational actors'. The result, according to him, is 'three general patterns of behavior', namely 'fear, self-help, and power maximization'.[6]

In addition, Mearsheimer follows Herz in arguing that international anarchy creates the 'security dilemma', which 'reflects the basic logic of offensive realism', and whose essence is that 'the measures a state takes to increase its own security usually decrease the security of other states' so that 'ceaseless security competition ensues'.[7] Themselves inspired by Herz's work, Ken Booth and Nicholas Wheeler have written at length on the security dilemma, describing it as the widely shared sense of uncertainty regarding the motives and intentions of states that acts as the prevailing existential condition in international relations. States are thus presented with a two-level strategic predicament: i) the dilemma of interpretation, i.e. the unresolvable uncertainty regarding the motives and intentions of states, ii) the dilemma of response. This situation is driven by a combination of material and psychological phenomena—primarily the 'ambiguous symbolism' of weapons and the 'other minds' problem. As a result, a security paradox can develop, whereby actors who only sought to ensure their own security 'provoke through their words or actions an increase in mutual tension, resulting in less security all round'.[8] Despite their potentially catastrophic risks, nuclear weapons are thus often justified by NWS as being their ultimate insurance policy in an uncertain world.

As we shall see in our discussion below, different authors assign different levels of importance to the concept of the security dilemma and what states can do to manage its impact. For example, Mearsheimer argues that little can be done to ameliorate the security dilemma, primarily because of a state's inability to overcome the problems of anarchy and eliminate uncertainty, in addition to the fear that relative gains can be achieved by states cheating on their commitments. The author thus acknowledges that he is applying a Hobbesian analysis, based around 'the absence of central authority' to the international state system.[9] This raises a number of questions with relevance to the challenges posed by nuclear disarmament. For example, how and why have deep and complex civil societies evolved domestically to prevent civil war whilst global civil society is much less able to constrain power and prevent conflict? Moreover, would it be possible to create a central authority at the international level that could support the elimination of nuclear arsenals and maintain disarmament and/or a NWFW, and, if so, what would the causes and consequences of this be?

We shall address these questions below, suffice to note for now that Mearsheimer sees bipolarity and nuclear weapons as having provided peace and stability during the Cold War, following which the US has, at times, acted as a 'pacifier' and 'offshore balancer' in Europe.[10] Importantly, Mearsheimer also notes that the internal dynamics and stability of anarchy aren't fixed, but vary according to changes in the balance of power, including the presence of nuclear weapons, and the number of great powers. However, he goes on to admit that a central problem his argument faces is that it is 'impossible to determine the relative influence of bipolarity and nuclear weapons' in establishing 'the absence of great-power war in Europe between 1945 and 1990'.[11]

Moreover, he makes the wider point that there is no evidence available regarding 'the effects of bipolarity and multipolarity on the likelihood of war in the absence of nuclear weapons'.[12] The caveat introduced here is important as Mearsheimer is admitting that it is not possible to be certain of the degree to which nuclear weapons contribute to peace and stability. Leaving aside the fact that the Cold War was a time of immense nuclear dangers, as shown by the Cuban Missile Crisis, which several prominent statespeople now accept was the moment of greatest peril for human civilization, Mearsheimer's admission raises the question of what other, unnamed factors, including actions by state and non-state actors, may have been involved in preventing great power conflict during the Cold War, and what lessons may be learnt from this—an issue I revisit in subsequent chapters.[13]

In addition to advancing his own offensive realist worldview, Mearsheimer critiques what he sees as rival positions and ideas, such as defensive realism and neoliberal institutionalism. For example, he claims that he developed his approach in response to the defects of defensive realism, which he argues is a 'good normative theory' but 'not a good descriptive theory', disagreeing in particular with Waltz's view that states 'should not maximize their power'.[14] As for neoliberal institutionalism, regarding the likelihood of reconciling the great powers, Mearsheimer argues that, whilst 'adversaries can cooperate, and that adversarial relationships can be transformed into friendly ones', cooperation is 'sometimes difficult to achieve and always difficult to sustain'.[15] This is because cooperation is inhibited by 'considerations about relative gains and concern about cheating'.[16] Moreover, great powers are unable to commit to the 'pursuit of a peaceful world order' because 'states are unlikely to agree on a general formula for bolstering peace' and 'policymakers are unable to agree on how to create a stable world'.[17] Indeed, as we shall see, such pessimism regarding the prevalence of competition and the difficulty of cooperation plays a central role in much of realist thought's concerns about—and objections to—nuclear disarmament.

Turning to Mearsheimer's views on the nuclear revolution, he claims that the existence of nuclear weapons ensures global peace, with Mutual Assured Destruction (MAD) making for a 'highly stable' world, helping to 'alleviate the vexed problem of miscalculation by leaving little doubt about the relative power of states'.[18] Other benefits of nuclear possession highlighted in his work include these weapons' ability to make states more cautious in their behaviour and 'dampen nationalism' by shifting 'the basis of military power away from mass armies and toward smaller, high-technology organizations'.[19] Yet whilst Mearsheimer argues that 'there is no question that nuclear weapons significantly reduce the likelihood of great-power war', he also admits that 'war between nuclear-armed great powers is still a serious possibility'.[20]

Despite this, and as we shall see with other authors who are either supportive of—or ambivalent about—the nuclear revolution, Mearsheimer values nuclear weapons so highly because he sees them as uniquely valuable in maintaining the necessary conditions for human survival. Given this belief, it is hardly surprising that he is strongly sceptical about the possibility or desirability of nuclear disarmament, noting that 'there is little evidence that world disarmament is in sight', so that he does not think it fit to explore the conditions and indicators of disarmament in any depth.[21]

As well as being a pessimist, one of the other principal reasons for Mearsheimer's nuclear enthusiasm and disdain for disarmament is that he is a convinced nationalist. Mearsheimer has boldly promoted this stance, noting in one interview that,

> speaking as an American, there would be only one state with nuclear weapons in an ideal world—the United States. Thus, if I could easily take away every other state's nuclear weapons and nip the Iranian and Iraqi nuclear programmes in the bud I would do so without hesitation.[22]

He has also argued that the US has and should aspire to nuclear superiority as this will 'likely' make it more secure.[23] Yet, according to his wider understanding of the conditions for international peace and stability, the US should remain an 'offshore balancer, not the world's sheriff', and thus a regional rather than global hegemon, because balanced multipolarity—as an international system—is much more conducive to the avoidance of war.[24]

Further evidence for the reasoning behind Mearsheimer's opposition to disarmament can be seen in such statements as 'a nuclear-free Europe has the distinction of being the most dangerous among the envisionable post-Cold War orders', with the danger here being that 'the Soviet Union and a unified Germany would likely be the most powerful states in a nuclear-free Europe'.[25] Mearsheimer thus advocates the 'carefully managed proliferation of nuclear weapons in Europe', also noting that he is not as 'sanguine' about the spread of nuclear weapons as Waltz.[26] Moreover, soon after the end of the Cold War he argued for Germany alone to acquire the bomb alongside the US maintaining its continental presence as the 'pacifier' that 'maintains order' and thus 'peace'.[27] It is clear here then that the continuation of a US-led security order, and its efforts to pursue goals beyond state survival are, for this author, acceptable so long as they focus on preventing the emergence of 'a potential hegemon in Asia or Europe' that could rival the US and upset the balance of power.[28]

Mearsheimer's focus on the prudent maximisation of US power may be contrasted with the more universalist viewpoint proposed by scholars such as Herz.[29] For example, Herz argued for a response to the nuclear revolution that moved on from 'particular national interest' towards an approach based on a 'universalist ... world-embracing feeling of responsibility'. Thus, whilst he strongly believed that the security dilemma 'has never before asserted itself more poignantly' than in the 'bipolar and nuclear world' of the Cold War, he equally strongly believed that a 'realist liberal, "universalist" solution of world problems' was the alternative to power politics and 'literally a matter of life or death'.[30] Yet Herz also asserted that as long as 'effective nuclear disarmament'—which would 'ultimately' be necessary—was not achieved, nuclear deterrence (with a general no-first-use policy) must be relied on 'to prevent a nuclear holocaust'.[31] Whilst we may disagree with elements of this last point, we may note that a normative approach based on such 'moral-political universalism' is surely more conducive to an investigation of the goals, indicators and conditions of NWS nuclear disarmament than one based on Mearsheimer's narrow nationalism.[32]

Notably, the focus on the international state system in Mearsheimer's work is such that there is an absence of substantial domestic analysis, which he puts down to the superior importance of structural factors on state's decision-making calculus.[33] Secondly, he is dismissive of the value of basing policy in this area on the popular will, because 'public opinion on national-security issues is notoriously fickle and responsive to manipulation by elites as well as to changes in the international environment'.[34] Yet, as with some of the other authors we will discuss below, Mearsheimer (with Russell Hardin) does recognise both how domestic politics can affect state's international behaviour and the impact of nuclear possession on domestic political processes. Exhibiting a keenness for consistency in his argumentation, he notes that the problem of nuclear possession harming democratic politics would not disappear in a NWFW if 'more powerful' conventional weapons were built to substitute for nuclear weapons because a substantial security establishment would still be required.[35] This observation is useful in that it again highlights the fact that nuclear possession, and military secrecy more widely, poses significant challenges to domestic liberty and democracy that need to be discussed in relation to disarmament.

Additionally, whilst Mearsheimer credits democratic peace theory (which, he observes, is based on the argument that 'democracies are more peaceful than non-democracies, regardless of the regime type of their adversary') as posing 'among the strongest' challenges to realism, he argues that 'it has serious problems that ultimately make it unconvincing'.[36] This is, for him, both because the theory is poorly evidenced and that 'stronger evidence exists for the claim that the pacific effects of democracy are limited to relations between democratic states'.[37]

Moreover, he argues that even if an authoritarian and illiberal great power (such as China) moved in a democratic direction, this would not ensure a pacifistic, 'status quo' attitude to international relations because the demands of the anarchic international system, which determines all states' behaviour, no matter the regime type, would remain in place.[38] Elsewhere, several scholars have added to the critique of democratic peace theory by pointing to evidence from the nuclear age suggesting that authoritarian great powers do not go to war with one another.[39] An approach to disarmament drawing on democratic ideas and ideals, as this study aims to do, would therefore benefit both from explaining how it relates to theories of democratic peace and by responding to the claim that progressive internal political changes will be superseded by the external pressures of anarchy in determining the character of a state's international behaviour. I therefore explore these issues further in section 2.7.

## 2.3 Defensive realism: Charles Glaser and Robert Jervis

An alternative to the pessimism of offensive realism has been developed by authors such as Charles Glaser and Robert Jervis and is often termed defensive realism. This approach agrees that anarchy is the primary characteristic of the international state system and a highly significant—though, importantly, not the sole—driver of great power behaviour in that arena. For example, defensive realism gives much more analytical space than offensive realism to the role domestic politics plays in state decision-making. In addition, whilst this approach accepts that anarchy encourages conflictual behaviour, it also emphasises that it prevents states from achieving goals that are in their 'common interest'.[40] For Jervis, the difference between the two realisms is thus that, whilst offensive realists 'see aggression and expansionism as omnipresent' or 'believe that security requires expansion', defensive realists believe that

> much of international politics is a Prisoners' Dilemma or a more complex security dilemma. The desire to gain mixes with the need for protection; much of statecraft consists of structuring situations so that states can maximize their common interests.[41]

The key point Jervis identifies here in relation to the security dilemma is that it may pertain in some but not all situations and can vary in intensity, being more or less 'vicious', depending on the circumstances.[42] For Glaser meanwhile, defensive realism has identified variables that cause the 'variation in countries' behaviour'.[43] In *The Security Dilemma Revisited*, he draws on Jervis's work to explain 'how the magnitude and nature of the security dilemma' depends on 'two variables', namely 'the offense-defense balance and offense-defense differentiation'.[44] Importantly, a world where it is possible to differentiate 'between offensive and defensive systems' allows, for Jervis, 'a way out of the security dilemma'.[45] Yet in the past, as he also notes, the difficulty of doing this has meant that arms control treaties 'have been rare', something that is also partly due to states 'not always' being 'willing to guarantee the security of others'.[46]

In addition, Glaser, who prefers the label 'contingent realist' for his ideas, given that the availability of security for states depends on 'empirical assessments of the offense-defense balance', notes that information sharing is important to help distinguish between 'greedy states', i.e. those with 'motives beyond security' and 'security seekers'. According to this analysis, 'the magnitude of the security dilemma' is influenced by the 'extent of the adversary's greed ... and of the adversary's unit-level knowledge of the state's motives'. Glaser develops this point by proposing that, in cases where 'democracies are believed not to have greedy motives', a democracy engaged in a 'military buildup' won't be seen to be as threatening as an authoritarian state acting in the same way. The result for him is thus that 'the democracy faces a less severe security dilemma; and interactions between democracies could result in a democratic peace instead of intense competition'.[47]

The idea of defence–offence differentiation also informs Jervis's argument that the nuclear revolution 'sapped' the driving force of the security dilemma—and was thus of major historical importance—by making defence central and overcoming the supremacy of offence.[48] More precisely, Jervis sees the advent of nuclear weapons as marking a shift from 'defense to deterrence', in the context of which, 'offensive weapons are those that provide defense'.[49] The expansionist ambitions of greedy great powers therefore became impossible given the dangers of MAD. For him, the 'high cost of war', which nuclear weapons help ensure, has thus today contributed to the creation of a security community 'among the leading states', i.e. the US, Western Europe and Japan.[50] Moreover, Jervis claims that nuclear strategy cannot work as, despite the fact that 'at best' nuclear possession 'will keep the nuclear peace', it will 'not prevent and, indeed, may even facilitate—the use of lower levels of violence', so that 'military victory is impossible'.[51] According to this logic, MAD may therefore weaken the security dilemma at the level of nuclear weaponry, but not necessarily at the level of conventional weaponry, depending on the policies pursued.

Despite the comments outlined, Jervis has also stated that he is 'deeply ambivalent' about the costs and benefits of the nuclear revolution. He therefore notes that it has brought both 'great security and insecurity', because 'on the one hand, mutual second-strike capability means that a major war is extremely unlikely; on the other hand, it means that if such a conflict should erupt, it is likely to destroy our civilization'.[52] This raises the question of what policies will prevent such a conflict erupting? In response, Jervis outlines several scenarios, one of which 'in the nuclear era' is where the superpowers rely on invulnerable submarine launched ballistic missiles, anti-submarine weapon technology is 'not up to its task' and 'limited nuclear options' are 'not taken seriously'.[53] Despite such scenarios promising stability and peace, nuclear conflict remains possible, with the 'basic question' of whether the nuclear revolution enhances security or not remaining 'unanswered, if not unanswerable' so that, for Jervis, 'it is not surprising that so many arguments rage'.[54]

Moving on to the question of nuclear disarmament, as with Herz writing in the 1950s, Jervis writing in the 1980s could not see the development of the political conditions supportive of nuclear disarmament in the near future. The priority for the latter during the Cold War was thus a restrained security approach to ensure results that were mutually acceptable for the great powers.[55] Moreover, in terms of precisely why meaningful disarmament processes are gridlocked, Jervis makes the important observation that 'to the extent that America's major interest is in preserving the status quo, nuclear weapons have brought the United States a real, although nonmilitary, victory'.[56] Owing to the fact that the US has security concerns beyond its national territory—which, according to Glaser's definition, would make it a 'greedy' state—'American resolve and will' come to be seen as 'primary'. The consequence of this for arms control agreements, according to Jervis, is that they have been treated by some in the US as 'bad ... not because they produce a less favorable military balance than would otherwise result, but because they produce psychological demobilization and disarmament'.[57]

These points complement Mearsheimer's nationalistic agenda by suggesting that the nuclear status quo supports Pax Americana. This logically means both that nuclear disarmament would jeopardise such a 'victory' and explains why those that support US state power and believe its security requirements include regional and/or global hegemony oppose nuclear abolition. However, the offensive realist claim that the 'peace and stability' of the Cold War was principally ensured by MAD and bipolarity alone is challenged by defensive realism, which, in Jervis's work, emphasises the diplomatic cooperation built between the Soviet Union and the US, alongside other factors such as 'the increased pain of war' given the 'political and economic modernization' following WW2.[58]

Whilst pessimistic regarding the prospects of nuclear disarmament, defensive realism is thus more optimistic than offensive realism when it comes to international cooperation, promoting the idea that uncertainty can be overcome if the incentives and motivations of states and their leaders are understood. The possibility of sharing information in this way, it is argued, allows decision-makers to move beyond escalating threats (as per deterrence) and arms races, in order to focus on compromises, reassurances and rewards. Furthermore, statespersons are encouraged to use diplomatic and political tools in order to develop transparency and empathy and signal their intentions. As discussed above, for Glaser, regime type matters here since democracies are more likely than authoritarian states to be seen as security seekers, thus weakening the security dilemma. Ultimately this all means that, with prudent leadership and compatible political institutions, states can exercise mutual restraint as a means of preventing conflict and, particularly in the current age, nuclear war.

As well as highlighting divergent views within realism, Jervis has sought to explain the differences between realists and neoliberal institutionalists. In doing so, he proposes that the two schools' 'disagreement over conflict is not about its extent but about whether it is unnecessary, given states' goals'.[59] Neoliberalism thus 'does not see more cooperation than does realism; rather, neoliberalism believes that there is much more unrealized or potential cooperation than does realism', which can 'at least in part' be explained by the fact that 'they study different worlds'.[60] Drawing on what he sees as the analytical strengths of neoliberalism, Jervis thus notes that,

> perhaps the most important path by which institutions can change preferences is through domestic politics. Drawing on liberalism, neoliberalism holds that states are not all alike and that preferences in part arise internally. To the extent that this is correct, international arrangements can alter the power, beliefs, and goals of groups in society in ways that will affect foreign relations.[61]

Whilst Jervis here accepts the neoliberal focus on how international institutions affect a state's internal politics and thus their foreign policy, he does not himself analyse the workings of domestic regimes, though he does note that other scholars have explored how 'the shape of domestic institutions affects both the chance of international agreement and the distribution of the benefits'.[62]

As with the 'different worlds' which realism and neoliberalism study, the domestic political sphere may thus be seen as *another world* which defensive realism is aware of but does not study in any significant depth. The implications of this will be seen with our subsequent discussion of nuclear possession and disarmament, given that theory in this area needs to be constructed from an analysis of both domestic and international sources of state behaviour. For example, we need to better understand how a state's relative strategic power and regime type affects its international goals—and ability to achieve them—in relation to nuclear issues.

In addition to defensive realism's interest in the neoliberal emphasis on institutions, Jervis and Glaser are very sympathetic to democratic peace theory. Jervis explains his enthusiasm by arguing that democracies with strong institutions are 'very likely to remain at peace with each other and to cooperate more readily than is true for autocracies or revolutionary regimes'.[63] Elsewhere, the same author notes that,

> if arms are positively valued because of pressures from a military-industrial complex, it will be especially hard for status-quo powers to cooperate. By contrast, the security dilemma will not operate as strongly

when pressing domestic concerns increase the opportunity costs of armaments.[64]

This is a rare case of Jervis outlining a domestic variable, i.e. whereby a military-industrial complex or other political 'concern' exerts an enabling or restraining influence over a state's defence and foreign policy, thus making it a source of international conflict or cooperation. In this case, the author identifies the state as a 'status quo power', which, based on his other comments, would probably make it a democracy.[65] If we accept this analysis, it is logical to extend it and propose that there may be lower or higher levels of democratic practices and processes within a state and that these may influence and correspond with the type of behaviour, for example cooperative or non-cooperative, that a state engages in on an international level, depending on the type of issue involved. Moreover, it seems reasonable to suggest that identifying the domestic political actors that enable or restrain action supportive of nuclear disarmament may both reveal to us important sources of a state's international behaviour as well as helping us understand the internal political changes that may be required in order for a state to become a greater champion of disarmament.

Turning now to the specific question of how defensive realism approaches nuclear disarmament, Glaser has done a significant amount of work imagining what this might entail, including its political and military requirements. Notably, as with other US-based realists we have encountered, his starting point for opposing the 'flawed case for nuclear disarmament' is whether this initiative will 'enhance' US security.[66] The other key question for Glaser concerns in what 'world' nuclear war would be more likely, to which he concludes that 'disarmament would not reduce the probability of nuclear war, so it would not provide what is commonly understood to be its key benefit.'[67] Overall, whilst like Jervis Glaser is sceptical regarding the benefits of the nuclear revolution given the risks involved, his reading of the literature on cooperation under anarchy leads him to conclude that the prospects for disarmament are 'extremely poor' and that there is also a 'variety of imposing domestic political barriers' that need to be overcome, not least for the US.[68] Moreover, he argues that there are significant dangers in shifting from what he sees as the current stability in the international state system—where the security dilemma is strongly mitigated by the war preventing effects of MAD and responsible arms control is in place, facilitating reductions towards small nuclear forces.

Overall, Glaser may therefore be said to represent a school of thought where national security—particularly, in his case, for the US—must come before disarmament, which corresponds with what may be termed a *security first* approach to eliminating nuclear weapons. According to this view, arms control measures may also, if further developed, make nuclear disarmament possible, if not very likely. Glaser thus critiques the notion that we should move to—what he claims would be—an unpredictable and potentially very volatile disarmed world that would 'reinforce' a spiral down in relations, where the possibility of accidental or deliberate nuclear use would persist, the potential for proliferation would become 'far more threatening' than today and where 'states' security would be very sensitive to cheating'.[69]

Leading arms control theorist Thomas Schelling has made several similar points, beginning with the argument that nuclear (and other) weapons cannot be disinvented. The main question he poses is, 'why should we expect a world without nuclear weapons to be safer than one with (some) nuclear weapons?'[70] For him, a NWFW cannot be one in which the great powers see 'every crisis' as a 'nuclear crisis' so that 'any war could become a nuclear war'.[71] This could occur because at the 'outset' of a 'major war', or when this seemed like an 'imminent possibility', those leading 'responsible governments' would race to build the bomb. The first states to acquire a bomb could then respond in several different ways, including using nuclear strikes to destroy their opponent's nuclear arms facilities 'in the interest of self-defense'.[72] Elsewhere, Schelling discusses other problems associated with 'total disarmament', arguing that 'some form of world government' would be required to 'police the world' and ensure against 'war and rearmament'.[73] The 'monopoly' of military power in such a centralised world authority would then have to find a way to 'improve' and 'stabilise' deterrence so that there is a 'balance of prudence', whereby states see maintaining a NWFW as preferable to rearmament.[74]

Jervis complements Glaser's and Schelling's view of disarmament, asking what will replace nuclear weapons in terms of deterrence? For Jervis, the important thing about nuclear weapons is the 'political effects' that they produce, 'not the physics and chemistry of the explosion' so that analysts therefore need to 'determine what these effects are, how they are produced, and whether modern conventional weapons would replicate them'.[75] Where Glaser's analysis differs from Schelling's, in particular, comes in the former's focus on the importance of domestic political barriers to nuclear disarmament. Glaser, in a way similar to Jervis and, for that matter, Herz, thus discusses the role of the military-industrial complex and other 'powerful interest groups' in promoting conflict, yet does this without going into much detail about these groups and how they function—an important gap in his thinking and the defensive realist project generally.[76]

Moreover, Glaser argues that the flaws with the case for nuclear disarmament mean that it is preferable to discuss how a 'permanent revolution' in international relations may be achieved, in order to ensure 'absolute long-term safety from the use of nuclear weapons'.[77] Echoing Schelling, Glaser also points out that any disarmament agreement would need to deal with the problem of rearmament given 'the coercive potential of a state gaining a lead in a rearmament race' so that 'states would have to coordinate their potential for nuclear rearmament, including their nuclear energy facilities'.[78] This observation again highlights how several authors from the defensive realist tradition see global nuclear disarmament as encompassing a range of issues beyond the disposal of the weapons themselves, including: the role played by

influential domestic political groups, for example, those pushing for the reproduction of the bomb; the potential need for world government; the recreation
of deterrence in a NWFW to ensure peace and stability; and the international
management of atomic power.

## 2.4  Structural realism: Kenneth Waltz

As with the other realist ideas discussed above, Kenneth Waltz's theoretical
approach to international relations (generally referred to as neorealism or
structural realism) centres on the contention that state behaviour is driven by
the demands of the anarchic international political system, which he describes
as 'politics in the absence of government'.[79] In this system, which is one of
'self-help', Waltz's structural realism posits that the main priority of states is
'to maintain their positions in the system', whereas offensive realism, as seen
in Mearsheimer's work, underscores the importance of power maximisation,
so that 'the international system provides great powers with good reasons to
act offensively to gain power'.[80]

For Waltz, this system has a structure formed by 'like units', i.e. states,
acting together, so that it is 'individualist in origin, spontaneously generated
and unintended'.[81] Importantly, the system operates in such a way that
'structures and units interact and affect each other'.[82] Whilst states within
the system are the same in terms of being 'autonomous political units', and
perform similar functions, they are also 'differently placed by their power' so
that 'great powers' have an increased ability to perform similar actions.[83] For
example, following the nuclear revolution, 'states relate to one another differently, yet each state still has to take care of itself as best it can', with nuclear
possessors, so Waltz claims, using their weapons 'in the service of peace' via
deterrence.[84]

For Randall Schweller, 'insecurity'—in Waltz's thinking—is therefore
driven not by 'greedy actors but by the inescapable self-help nature of the
system', so that states seek maximum security rather than maximum power,
as proposed by offensive realism.[85] This, for Schweller, as well as Glaser and
Jervis, is an error which must be corrected by bringing 'the revisionist state'—
meaning those which have non-security based, expansionist goals—'back in'
so that 'differences in state goals' are studied alongside 'anarchy and the distribution of capabilities'.[86] Waltz does note that states have 'endlessly varied'
goals beyond survival, and does differentiate between states according to
national power, for example, by discussing at length the US's dominance in
his work and how its behaviour affects the threat perceptions of other states.[87]
However, as with other US-based analysts utilising realist ideas that we have
encountered, Waltz is keen to present the US as a 'benign' manager of the
international state system, and as a 'liberal dominant power'.[88] As we shall
see, the 'status quo bias' which Schweller claims Waltz commits when he fails
to distinguish between 'greedy' and 'security-seeking states', helps Waltz to

*Increases risk of accidental use*

sustain the belief that when it comes to the spread of nuclear weapons 'more may be better' and nuclear disarmament is neither desirable or realisable.[89]

For Waltz, the character and nature of anarchy varies according to the number of great powers in the system—and thus whether the system is unipolar, bipolar or multipolar.[90] As with Mearsheimer, Waltz believes that bipolar relations are the most stable of the alternative systems so that over the course of the Cold War 'the longest peace yet known rested on two pillars: bipolarity and nuclear weapons'.[91] After the Cold War, the world moved away from bipolarity to unipolarity as the US became the world's only superpower. Now, however, Washington's power is being challenged and multipolarity is 'developing before our eyes'. For Waltz, this is entirely natural given how the 'balance of power' operates, so that 'some states try to increase their own strength' or 'ally with others to bring the international distribution of power into balance' in an effort to ensure their autonomy and survival.[92]

In terms of Waltz's response to neoliberal institutionalism, as with offensive realism, anarchy is perceived as making 'collective action for the common good hard to achieve', in addition to problems such as multilateral coordination and free riding.[93] Moreover, if great power cooperation is to be viable, the US needs to lead the way, given that 'international institutions are created by the more powerful states, and the institutions survive in their original form as long as they serve the major interests of their creators, or are thought to do so'.[94] As for democratic peace theory's potential contribution to world peace, Waltz raises several concerns. For example, according to him, proponents of this theory argue that only 'democracies of the right sort' can bolster the cause of peace, yet 'democratic states, like others, have interests and experience conflicts'.[95] A state may thus try to become a democracy and live in peace, but so long as the international environment remains the same, i.e. anarchic, this won't count for much in terms of the state's international actions. Furthermore, the 'external behavior' of 'dominant powers', for Waltz, 'bears little relation to their internal political composition'.[96] Democracies may therefore exist in peace with one another, but even if 'all states became democratic, the structure of international politics would remain anarchic' meaning that the threat of war would remain.[97]

Jervis has responded to Waltz's arguments by positing that he ignores 'regime type' and 'narrow domestic interests and how they are aggregated through domestic institutions'.[98] Jervis's argument here is that we need to consider how a state's internal politics and regime vary if we want to explain international behaviour, which corresponds with Schweller's point about the need for a differential analysis of a state's goals. Yet if we look more closely at Waltz's work, we can see that, in several places, he does note the important role domestic politics plays, stating that 'the causes of war lie not simply in states or in the state system; they are found in both'.[99] The issue here, as with Mearsheimer, is which factor—agent or structure—plays a bigger role. For example, Waltz notes that 'small internal changes' cannot compare to the importance of the 'international structure', so that 'external behaviour bears little relation to … internal political composition'.[100] Elsewhere he also identifies how 'internal military and political pressures' can drive arms races between nations.[101] We may conclude from such statements that Waltz believes that changes to international behaviour and order within the current system is possible and, to some extent, driven by domestic developments, whereas changing the structure of the system itself is far more difficult.

This aspect of Waltz's work is highlighted by Campbell Craig who notes how 'Waltz conceded that unit-level phenomena affect international politics in three ways', for example: i) ideas can have an impact on 'questions of great power war and peace', ii) national leaders can manage systemic stability, iii) fear of nuclear war engenders caution by decision-makers.[102] This discussion is important for the purposes of this study because of our need to identify drivers of the requisite changes to domestic and international politics (which may, in the latter case, include a transition from anarchy to a different type of structure) supporting nuclear disarmament action of various scales. In order to begin doing this and consider how Waltz's views on unit-level behaviour relate to questions of nuclear disarmament, it is first necessary to review his justification for nuclear possession and his arguments in favour of the spread of nuclear weapons.

The arrival of nuclear weapons at the end of WW2 is presented by Waltz as one of the 'greatest … within-system changes' to take place 'in modern history, or perhaps in all of history'.[103] The invention of the bomb was thus epoch-making but did not alter the anarchic structure of the international political system itself. Moreover, for Waltz, since structural forces lead to the logic of self-help and power politics, great powers value nuclear weapons because they secure their vital interests and uphold the international order. According to this widely held view, highlighted by Keir Lieber and Daryl Press, the 'military stalemate' of MAD, which has existed between the US and the USSR/Russia since the 1960s, 'made the world relatively stable and peaceful' by inducing 'great caution in international politics', discouraging 'the use of nuclear threats to resolve disputes' and generally restraining 'the superpowers' behavior'.[104]

Waltz presents his defence of continued nuclear possession in terms that are, in several ways, similar to the arguments already outlined by Mearsheimer, Glaser and Jervis. First, nuclear weapons are, Waltz argues, 'useful' for great powers in dealing with the security dilemma because they 'make the waging of war among them unlikely'.[105] Secondly, nuclear weapons are low cost compared to large-scale conventional forces. Thirdly, these weapons are only needed for deterrence—meaning a secure second-strike capability—because they are 'useless for fighting wars and even for threatening blackmail'.[106] Fourthly, nuclear weapons are the 'only peacekeeping weapons the world has ever known', being 'inherently stable' given that they 'induce caution' in their possessors.[107]

Waltz has also embraced the gradual spread of a nuclear-armed crowd, celebrating nuclear weapons for lessening the 'intensity as well as the frequency of war among their possessors'.[108] Moreover, his argument for the spread of nuclear weapons considers why states targeted by the US—such as Iran and North Korea—might seek the bomb, stating that 'conventional defense and deterrence strategies have historically proven ineffective against the United States, so, logically, nuclear weapons are the only weapons capable of dissuading the United States from working its will on other nations'.[109] Several scholars have critiqued Waltz's position on the spread of nuclear weapons. For example, T. V. Paul posits that Waltz's arguments do not correspond with political reality, noting that the number of countries that acquired nuclear weapons after the original five 'is so small that these cases seem more like an anomaly than the norm'.[110] Maria Rost Rublee also concludes that 'realism' tends to 'overpredict proliferation' and does not convincingly explain why so few states have developed nuclear weapons.[111]

With regard to the conditions and indicators of nuclear disarmament, whilst accepting the potential of arms control as a form of cooperation, Waltz has stated that 'it would be strange' for him to support the elimination of nuclear weapons 'as they have made wars all but impossible' and, in any case, 'it's impossible' to get rid of these weapons 'entirely'.[112] According to structural realist logic in general, non-proliferation, nuclear disarmament or rollback may occur if a threat is removed or following the state receiving security guarantees from the US. Thus, despite Waltz arguing that international efforts should 'concentrate more on making large arsenals safe' and less on preventing weaker states obtaining small nuclear forces, he also observes that 'we should be careful about conveying military threats to weak states' because the leaders of such states may seek the bomb owing to their 'fear' of losing power.[113]

Whilst being generally enthusiastic about the peacekeeping qualities of nuclear weapons, when properly managed, Waltz does admit (like other authors we have reviewed) that their possession contributes to authoritarian government and secrecy at home.[114] However, his treatment of this topic is brief and not presented as a problem requiring a solution—presumably because he believes such issues are a side effect of what he describes as states 'taking care of their own security'.[115] It also needs to be recognised that, rather than nuclear weapons providing a 'cheaper' strategy for NWS, former US Defense Secretary William Perry and retired British Vice Admiral Jeremy Blackham have noted that nuclear powers have sought strong—and thus very costly—conventional militaries to ensure the credibility of their nuclear arsenals, for example, in terms of escalatory threats moving from conventional to nuclear use.[116] In this sense, when estimating the costs of nuclear possession we need to take into account both the expense of maintaining sufficiently powerful military establishments and their domestic implications for liberty and democracy.

Elsewhere, Waltz engaged in a high-profile debate with Scott Sagan concerning the desirability of horizontal proliferation. Sagan countered Waltz's claim that more nuclear weapons states 'may be better' by arguing that such moves would be potentially destabilising, given the problems of command and control and the risks of 'deterrence failures and accidental use'.[117] Whilst Sagan argues that nuclear zero is the 'best option', Waltz countered that the recent 'vogue of abolition' is driven by experts in the US who are concerned that nuclear proliferation to 'the weak' will 'limit what the strong can do to them'.[118] Furthermore, as with Glaser, Waltz is concerned about the stability of a disarmed world, arguing that, with the disappearance of bipolarity, it is necessary to 'compare the problems of balancing in conventional and nuclear worlds'.[119] Waltz therefore argues that nuclear disarmament would raise the likelihood of conventional war, particularly given both the issue of miscalculation—which, he claims, is much higher in a conventional than a nuclear world—and the US's predilection for global dominance, so that the problems of war and how to balance rival military forces need to be solved before abolition can be contemplated.

In addition, Waltz contends that, even if a nuclear weapons ban could be agreed, it 'would be impossible to police and enforce' given the ease with which states could cheat on their commitments and become involved in a rearmament race.[120] Moreover, he posits that 'controlling and moving toward the elimination of nuclear weapons' would require 'nothing short of an unimaginably competent and despotic international regime', a scenario equating to a 'world tyranny' which, he argues, no one would want.[121] In Waltz's eyes, nuclear disarmament thus, ultimately, clashes with the fundamental goals of the great powers and would involve several highly undesirable consequences for global order and security.

## 2.5  Cooperation under anarchy: Robert Keohane and Nicholas Wheeler

In contrast to the pessimism of those whose realist perspective leads them to emphasise the prevalence of competition and conflict in international relations, a number of scholars take a more optimistic view. For example, Robert Keohane contends that, contrary to realism, conflict between states is not inevitable and that states can, 'when complementary interests exist', construct international institutions to alleviate security dilemmas and enable long-term cooperation to the benefit of all.[122] As Keohane explains, this approach may thus be distinguished from Waltz's structural realism 'by its emphasis on the effects of international institutions and practices on state behavior'.[123]

Keohane also describes how his work on cooperation and the design of institutions is 'infused' with normative intent, in order to solve 'the most pressing practical problems', including the need for humanity to avoid a nuclear conflagration. Reviewing the meaning of realism, he therefore asserts that it

> sometimes seems to imply, pessimistically, that order can be created only by hegemony. If the latter conclusion were correct ... at some time in the foreseeable future, global nuclear war would ensue ... No serious thinker could, therefore, be satisfied with Realism.[124]

His 'liberal institutionalist' theory sees international cooperation as requiring that 'the actions of separate individuals or organizations—which are not in pre-existent harmony—be brought into conformity with one another through a process of ... "policy coordination"'.[125] According to this approach, rather than international order being an inevitable result of power struggles, it may be deliberately—and more productively—shaped by states that are able to realise their potential to cooperate.

Whilst Keohane does not directly discuss the types of interstate relations conducive to disarmament, his argument that cooperation under anarchy is possible and necessary in the absence of world government supports the idea that states can together create an international environment that suits their mutual interests and which enables them to take disarmament action. Moreover, he explores problems related to nuclear issues when outlining the difficulties states face in constructing institutions in the 'military-security' arena. This area can be particularly challenging because of the 'high cost of punishing defections, the difficulties of monitoring behaviour and the stringent demands for information that can be imposed when successful defection can dramatically shorten the shadow of the future'.[126] The design of effective institutions therefore requires the systematic working out of 'how to get the incentives right in constructing institutions, and what scope global institutions should have'.[127] As a result, regimes may be created with 'rules' that 'may provide opportunities for governments to bind their successors, as well as to make other governments' policies more predictable'.[128] The uncertainty prevalent in military-security cooperation might therefore be overcome if the state's behaviour is altered to take into account the 'preferences of others' so that policy coordination is achieved between them.[129]

Such strategies could be applied to arms control and disarmament negotiations as well as the reformation and construction of institutions to both ensure mutually beneficial outcomes and consider how the political environment suits different types of initiatives. Indeed, Robert Axelrod and Keohane point out that a key benefit of 'international regimes' is that they 'reinforce and institutionalize' reciprocity, making 'defection' less acceptable and more costly.[130] Yet, as they also note, in a passage which has particular relevance for this study and its focus on *institutional democratisation*, 'arms control negotiations involve not merely bargaining between governments, but within societies as well'.[131]

The result is that 'political institutions' follow the 'preferences' of domestic actors by design, yet such institutions can, Keohane and Helen Milner argue, 'also have independent effects' because they 'create rules for decision making, help to structure agendas, and offer advantages to certain groups while disadvantaging others. Over time, strong institutions may even shape actors' policy preferences.'[132] The overall point is that national and international institutions each contain 'constraints' which 'interact' in varying ways according to prevailing 'political-economic conditions'.[133] Thus, 'international developments could affect the coalitions that form in domestic politics'—an observation which may be applied to the purposes of this study in several ways, by, for example, considering which developments may support an NWS moving towards *institutional democratisation* and/or nuclear disarmament—or not.[134]

Elsewhere a range of scholars, including Perkovich and Acton, Cortright and Väyrynen, Harald Müller, William Walker and Nicholas Wheeler have investigated how nuclear disarmament may be advanced through states cooperating, an approach I shall hereafter term *cooperation with disarmament*. Whilst the work of these authors is recognisably distinct, it is sufficiently compatible to be grouped together in this way. The main area of agreement across these and other authors is that the US, Russia and China must engage in geostrategic cooperation as part of a political process in which they move gradually to diminish their reliance on nuclear weapons.[135]

Of the recent works within the *cooperation with disarmament* category, Cortright and Väyrynen's *Towards Nuclear Zero* provides a detailed and thorough set of ideas and proposals. Of particular importance is their conclusion that nuclear disarmament must be understood fundamentally as a political process. This necessitates the resolution of conflict and the development of 'cooperative political and economic relations' to make NWS feel secure enough to disarm. For example, policies such as the 'institutionalisation of democracy, economic interdependence and participation in multilateral institutions' are proposed as 'steps in the direction of increasing the prospect and sustainability of disarmament'.[136] A similar series of proposals can be found in Perkovich and Acton's *Abolishing Nuclear Weapons*, which begins by asking how the 'security conditions which would permit nuclear weapons to be safely prohibited' may be created, and how 'might measures to implement such a prohibition be verified and enforced?'[137] The authors address this question by discussing nuclear abolition as a 'co-evolutionary' process involving 'reciprocal step-by-step progress, in which nonproliferation and arms-reduction measures emerge from changed political and security environments and vice versa'.[138]

Differences occur amongst authors advocating *cooperation with disarmament* when it comes to the emphasis placed on the process by which a balance of interests—rather than a balance of power between the major states—might be achieved. For example, Wheeler emphasises trust-building between the great powers, whereas Perkovich and Acton focus on regional conflict resolution.[139]

These approaches go some way towards addressing the problems discussed in Chapter 1 regarding the technical requirements of nuclear disarmament. For example, Wheeler explores how the 'good relationships' Findlay identified as being vital for an effective verification and compliance regime in the creation and maintenance of a NWFW might be established. First, governments must base their security on mutual trust rather than mutual fear—as exemplified by the possession of nuclear weapons. Contrary to the 'standard contention', Wheeler argues that states can do this by using mutual cooperation and trust-building measures to move away from competitive or individualistic conceptions of security towards more cooperative systems within the anarchic international system.[140] States may thus, he posits, mitigate uncertainty and build initiatives supportive of nuclear non-proliferation and disarmament efforts.

As we have seen in our discussion of Glaser and Schelling's work, a key concern here is that, even if it were possible to reach zero in an atmosphere of fear and distrust, a non-nuclear world could be a far more dangerous place to live than our current one. This is because of the problem of 'hedging', whereby a state might secretly maintain or acquire the bomb, out of fear either that others were doing the same, or that they might do so in the future. Wheeler therefore argues that global nuclear disarmament can lead to a more secure world, but on the critical condition that 'each step on the road to "global zero" is conceived as a process of trust-building'.[141]

The first problem that needs to be raised concerning the authors I identify with the *cooperation with disarmament* approach, is that they generally do not consider how nuclear disarmament could take several different shapes and forms. For example, disarmament may be unilateral, bilateral or multilateral and therefore require lower or higher levels of cooperation. A nuclear disarmament process may also involve different degrees of cooperation depending on the possessor state(s) involved and the material-political context, for example, whether a higher level of disarmament is required, which relates to the size and scale of the nuclear weapons complex of the state in question.

A second concern is that *cooperation with disarmament* does not fully take into account domestic sources of political behaviour and cooperation. For example, a state may be 'greedy'—and thus less cooperative, or more moderate—and thus more cooperative. A theory of nuclear disarmament should therefore take a state's strategic goals and behaviour into account when assessing how cooperative it is or is not being, regarding an international proposal or agreement (for example, on arms control), alongside assessing the aims and influence of the main domestic actors driving a state's behaviour.

Connected to this is the third concern that *cooperation with disarmament*—and similar studies that adopt a liberal viewpoint—focus on action pursuant to disarmament being taken by existing decision-making elites. For example, Wheeler argues that 'achieving radical reductions in nuclear arms' will require NWS leaders with both 'imagination' and 'empathy' that can be translated into 'state policies that can build trust', and also 'the domestic political support to take a series of unilateral measures'.[142] Yet whilst these attributes may be vital, studies in the *cooperation with disarmament* category tend not to examine whether certain types of elite actors—for example, in the NWS— are actually inherently incapable of cooperating and building trust on this topic, given their deep ideological investment in nuclear possession. If this is accepted, an investigation of what more radical domestic political change is necessary in NWS for disarmament to advance is also required.

In relation to this, Lawrence Freedman and Jacob Nebel have made important criticisms of Perkovich and Acton's analysis. For example, Freedman, reviewing both the approach of these authors and that of similar investigations, questions how far nuclear disarmament can 'be taken as an elite project?'[143] He goes on to pointedly characterise the style of such works as 'geopolitical engineering enterprises', whereby 'barriers are to be cleared by judicious treaty language here, a technical fix there, and a confidence-building measure to follow'. The key question for Freedman here is thus where public opinion fits into nuclear disarmament given that in these works it

> appears rather distant, as nothing more than a supposedly approving chorus. Yet governments must be accountable to their electorates. If this undertaking is going to be treated with the seriousness it deserves over an extended period, public opinion will need to be engaged.[144]

Similarly, Nebel chides Perkovich and Acton for 'leaving popular opinion out of the picture', citing the authors' own framing of their work as part of an 'enormous renovation project' for 'experts from a representative range of states' as a source of what he terms 'movement pessimism'.[145] Nebel claims that this is a serious error because, with regards to the US, 'the historical record indicates that the disarmament movement has immense political potential. The disarmament movement has constrained decision-making even in the most pro-nuclear-weapons administrations.'[146] The key question Freedman and Nebel's criticism raises concerns what *theory of political change* is being proposed by the different approaches to nuclear disarmament we have reviewed. This study therefore develops existing studies emphasising the need for international cooperation to achieve nuclear disarmament by investigating both the role decision-making elites across NWS play in helping or hindering such action and the role that popular movements can play in advancing progress in this area.

Notably, Wheeler does discuss domestic political change in relation to how Argentina and Brazil reversed their nuclear rivalry in the 1980s.[147] In addition to leaders from both sides developing empathy for one another's 'security concerns by taking a series of reciprocal confidence-building steps', Wheeler argues that another key 'lesson' to be learned from this experience is the 'importance of democratisation to trust-building', because leaders came to power 'who were aware of the growing political and economic costs and risks of pursuing a unilateral approach to security'.[148] Yet whilst it is argued that such findings, including regarding democratisation, could be applied 'for building trust elsewhere', this is not meant universally but for 'cases such as South Asia, Northeast Asia, and the Middle East'.[149] Yet, as I shall argue in more depth in later chapters, we need to understand the domestic politics of nuclear possession for *all* NWS—including Western democracies—and the ways in which democratisation may assist the path to national and international disarmament.

The fourth concern is that international cooperation can work against nuclear disarmament as much as it can support it. For example, as discussed in Chapter 1, the NPT was designed by the Soviet Union and US during the Cold War in a way which protected their nuclear status. Since then, the NWS have repeatedly stated that they are committed to achieving a NWFW on a multilateral basis, despite all the evidence to the contrary. The NWS maintain this façade as it is important to them that the NNWS do not lose faith in the NPT bargain, as *their* cooperation is essential to prevent nuclear proliferation and the loss of the exceptional influence that NWS currently enjoy. Whilst the great powers may therefore criticise one another's military build-ups and behaviour, there is also a tacit understanding that they must cooperate to prevent NNWS from delegitimising and diminishing their position in the international hierarchy of states, including through obstructing disarmament action.

This can be seen in the recent concerted efforts by NWS to boycott and block negotiations on the TPNW, which received the support of 122 nations in a vote at the United Nations in July 2017.[150] In a world consisting of what Joseph Nye and Keohane describe as 'asymmetrical interdependence', whereby 'the more asymmetry you have in your favor, the stronger you are, the more you have of some resource, the more your advantage in influencing the outcome of an event', uneven power relationships are continually generated between states.[151] The fact that the great powers, to different degrees, dominate international institutions that have ostensibly been built on equal cooperation and participation thus needs to be kept in mind when identifying which actors are responsible—and to what extent—for the present disarmament impasse.

Overall, *cooperation with disarmament,* as an approach that focuses on the challenges of securing cooperation under anarchy in order to achieve nuclear disarmament, is thus primarily important because of its focus on disarmament's international security implications and requirements. However, as Keohane and Milner suggest, it is also vital to appreciate the 'relative importance' of the 'constraints and incentives' in existing 'national institutions', and those 'imposed' at the international political level—as well as how these factors 'interact'—to understand the behavior of political decision-makers.[152] In terms of providing a full explanation of the causes and consequences of nuclear disarmament in and between NWS, this means examining relevant evidence in the domestic arena in tandem with evidence pertaining to the international arena. Ultimately therefore, given the absence of an analysis of the domestic politics of nuclear possession in and between

NWS, we must conclude that *cooperation with disarmament* makes several valuable contributions to the topic, but is, for our purposes, one that can only fill in some parts of the complex disarmament puzzle.

## 2.6  Moving beyond realism: Campbell Craig and Daniel Deudney

Just as there is a lively tradition of discussion within realism, and several proposals regarding how such analyses may be improved upon from liberal and institutional theorists, so there are a variety of critiques of it from without. For the purposes of this inquiry, Campbell Craig and Daniel Deudney's work provides two the most relevant responses worth reviewing. Craig, in his study titled *Glimmer of a New Leviathan: Total War in the Realism of Niebuhr, Morgenthau and Waltz*, is keen to present nuclear weapons and war as a problem not just for realism but for 'formal American thinking about international politics' as a whole.[153] He argues that, despite their importance, such concerns play a 'strikingly small part' in these works, so that Mearsheimer, amongst others, 'consigns the problem of nuclear war to the margins'. Craig explains this by noting that such authors maintain an obdurate belief that 'nuclear deterrence has largely eliminated the possibility of thermonuclear war'.[154] At the same time, Craig—with Sergey Radchenko—has recently written of the nuclear revolution's importance in preserving peace because 'the aversion of leaders to nuclear war' is what 'best explains the absence of major conflict among nuclear powers over the past 70 years'.[155]

In contrast to scholars who dismiss the 'implications of the thermonuclear revolution', Craig notes that Jervis and Deudney have been at the forefront of examining how nuclear weapons 'can be reconciled…with modern Realism'.[156] Given the unwillingness of mainstream scholarship to deal with these issues, Craig questions whether realism, which he casts as a 'static conception of international relations', can survive as a theoretical approach if it does not seek to advance past 'the continuation of anarchy and of nuclear deterrence'.[157] This is because, whilst nuclear deterrence may have worked to keep the peace in the Cold War, it will, he writes, 'sooner or later' result in 'thermonuclear war' that is likely to 'kill hundreds of millions of people and possibly exterminate the human race'.[158] Craig, like other scholars such as Nick Ritchie, therefore accuses defenders of the nuclear status quo as being 'utopian' in denying the inevitability of eventual nuclear conflict. For Ritchie, 'real' realism also logically entails investigating the conditions required for progressive action on nuclear disarmament in order to create a NWFW.[159]

Craig suggests that, if realism can be saved as a relevant theory, it will be because 'the fear of nuclear war', threatening human survival, could provide the motivation to overcome anarchy.[160] This may occur via 'common action' to eliminate nuclear dangers and involve 'the establishment of an authoritative, centralised world state'. For Craig, any meaningful answer to the threat of nuclear catastrophe can, therefore, 'only be attained by concerted international cooperation', yet he asserts that this is 'something very difficult to achieve in an anarchical interstate system'—a problem we have already discussed.[161] The development of a world state, which for Craig would entail 'the abolition of anarchical great-power politics', has, he points out, been hinted at by leading 'realist' thinkers such as Reinhold Niebuhr and Hans Morgenthau.[162]

However, in addition to Craig and Waltz, Niebuhr and Morgenthau have serious concerns about the consequences of creating such a state, principally because it might create a global tyranny. Craig discusses how, as with Herz, Morgenthau and Niebuhr found that a shift away from international anarchy towards world government required the development of a world community, with group loyalties shifting from the national to the international level.[163] He also raises the important point that Niebuhr, Morgenthau and Waltz all 'gravitated towards the normative goal of great power peace', a shift towards an internationalist rather than a nationalist position of a kind similar to Herz's, as previously noted.[164]

A problem that once again emerges from Craig's analysis of the three authors discussed in *Glimmer of a New Leviathan* is the lack of detailed thought given to the political forces that could drive nuclear disarmament, great power peace and a reformed international order which ameliorates or overcomes the problems posed by anarchy. Overall, this particular work therefore does not thus seek to provide answers to the methodological, normative and political issues raised by the thermonuclear revolution, but does highlight some of the key problems and questions that leading exponents of realist thought faced when coming to terms with it. For insights into the indicators and conditions for nuclear disarmament we must therefore look elsewhere in Craig's oeuvre.

For example, in his article *Weberian World Government in the Nuclear Age*, Craig argues that 'the unique qualities of the nuclear revolution' make 'securing a monopoly over violence' a task which is 'even more important to a prospective world government than it is to a traditional nation-state'.[165] Moreover, drawing from the thinking of Alexander Wendt, Craig writes that finding a means of compelling or persuading the US to join an emerging world government 'constitutes the single greatest obstacle' to its formation.[166] Meanwhile, in the book Craig co-authored with Radchenko, *The Atomic Bomb and the Origins of the Cold War*, it is argued that the possibility of a state cheating and building a nuclear weapon secretly means that 'international atomic control requires a qualitatively different level of international action if nations are to be persuaded to place their trust in it'.[167] The problem of cheating thus means that nuclear disarmament would require all nations to 'transfer all of their technological and physical means of building atomic weaponry to a powerful international agency' which would need to persuade all nations that it could 'permanently ... prevent any nation from secretly building another bomb'.[168]

The argument Craig and Radchenko present—that the choice the major powers face in the nuclear age is between 'sovereignty or international government'—is a formulation very similar to Waltz's view. The centralisation of power involved in disarmament thus, for these authors, essentially means that the agency responsible for managing nuclear knowledge and materials 'would have to become like a state itself'. This was something during the Cold War that neither the then Soviet Union nor the US was willing to countenance because it would, Craig and Radchenko assert, mean relinquishing their 'sovereignty to the agency'.[169] As Craig makes clear, when it comes to contemporary discussions of what world government might entail, Daniel Deudney has, in his work *Bounding Power,* provided 'the fullest and most creative vision yet of formal world government in our age'.[170] It is therefore appropriate to investigate Deudney's work further at this point, to build on the ideas and insights already discussed.

To begin with, Deudney directly tackles problems of institutional power in his work, taking an approach that is both descriptive and prescriptive. His ideas are also developed out of a critique of the prominent scholars we have encountered, including Waltz, Mearsheimer and Keohane. Whilst great power conflict did not occur during the Cold War, Deudney argues that 'it is impossible to say' whether this was the result of nuclear weapons or some other cause.[171] Alongside Jervis, he therefore believes that the central questions about the nuclear revolution—'How likely is deterrence failure? What will happen after nuclear use?'—are 'unanswerable with any assurance'.[172] Moreover, following the end of the Cold War, we have moved into the second nuclear age, which is defined by 'the diffusion of nuclear weapons capability to small and often revisionist states, and possibly also to non-state actors' so that the possibility of 'deterrence failure is much more likely'.[173]

Deudney moves on to pose the intriguing question of whether the dynamics prevalent in the current age 'will provide the catalytic impetus to a fuller revolution not just in the conduct of states but in the basic practices and structures of the anarchic interstate system itself'.[174] However, elsewhere, in an observation complementing Craig's view, Deudney states that nuclear weapons have created a 'paradoxically perverse effect', whereby they necessitate 'an exit from anarchy', given the fact that the system is prone to 'potentially catastrophic failure', but simultaneously 'impede the traditional path to exit from anarchy', by which he means, 'some form of authoritative world government'.[175]

In order to move beyond this problem, Deudney develops and adds the idea of 'negarchy', as a third type of international system, to anarchy and hierarchy.[176] Deudney's advocacy of negarchy is based on the need to find a way out of the contemporary international military-political situation, which combines anarchy and the pronounced 'violence interdependence' created by nuclear weaponry. Importantly, negarchy is 'self-ordering', meaning that it avoids the top down imposition of order in favour of an 'alternative "republican federal" set of practices' where actors are 'authoritatively ordered by relations of mutual restraint', rather than hierarchical subordination and anarchical orderlessness.[177] To bolster his position, Deudney employs the concept of Republican Security Theory, which is based on the principles of liberty, sovereignty and limited government, as a means of refining and improving on liberal and realist proposals for managing international order.[178]

As with some of the other authors promoting realist thought noted above, Deudney—in addition to discussing the meaning of the nuclear revolution for international politics—notes the impact that the possession of nuclear weapons has had on domestic liberty, but goes into greater depth about what this means in practice. For example, in *Bounding Power*, Deudney discusses how nuclear weapons are 'intrinsically despotic' and have created 'nuclear monarchies' in all nuclear-armed states. Deudney identifies three related reasons for this development:

> the speed of nuclear use decisions; the concentration of nuclear use decision into the hands of one individual; and the lack of accountability stemming from the inability of affected groups to have their interests represented at the moment of nuclear use.[179]

For Deudney, a state's moves to ensure its security, for example, via nuclear possession, can therefore lead to the centralisation of decision-making in powerful institutions, with deleterious results for democracy and liberty, so that he identifies a realist 'neglect of domestic hierarchy as a security threat'.[180] Conversely, 'republican and Liberal theory and practice' addresses this problem by focusing on 'the restraint of domestic hierarchy as a security threat over progressively larger spaces'.[181] The legitimacy of the constitutional republican government he proposes, created to ensure state security, would thus derive from 'popular sovereignty'—the will of the people.[182] In republics, the people are in control of military power 'either directly or indirectly', so that consensual, rather than coerced, arms control is particularly compatible with this system of government.[183]

Such thinking informs Deudney's response to what he describes as the 'nuclear political question', whereby he identifies four approaches that outline 'the relationship between nuclear weapons and the state system'. These he labels as 'classical nuclear one worldism, nuclear strategism, automatic deterrence statism' and 'institutional deterrence statism'.[184] Deudney then presents the approach he subscribes to, namely, 'federal-republican nuclear one worldism'.[185] This position, which is worth presenting in full, argues that nuclear weapons have

> rendered the statist approach to security nonviable, and that security in the nuclear era requires the establishment of an institutionalized division between territorial units and nuclear capability. Instead of either the continuation of an interstate anarchy or the establishment of a world state, a federal-republican union of strong mutual restraint is needed to provide security. This view holds that a world hierarchical government would

entail an uncheckable concentration of power, and is unnecessary in the absence of an interplanetary threat.[186]

Crucially, whilst this approach (and Deudney's overall analysis) provides us with the kind of two-level and normative approach required by nuclear disarmament, focused on domestic *and* international security and liberty, and how these levels interact, he does not advocate or present a theory of nuclear disarmament. Instead, because Deudney values how nuclear weapons 'have greatly reduced the problem of interstate aggression, while creating a new threat of general annihilation', he prefers to find a means of managing the power of nuclear weapons for the common good. His new approach thus proposes that the 'territorial state system' not be replaced but 'complemented with a nuclear containment and restraint system' so that 'nuclear capability is separated from state control and paralyzed'.[187]

In outlining 'federal-republican nuclear one worldism', Deudney—like the other authors working in the realist tradition discussed above—does not embark upon a detailed examination of the domestic or international nuclear politics of the major powers. Nor does he identify the potential political sources of change that will facilitate the transformations he outlines.[188] This may partly be because Deudney exhibits a similar caution to Mearsheimer, arguing that it is not possible to definitively prove his proposals or explanations. The question of what will drive nuclear restraint or disarmament and the extent to which these will be popular or elitist projects thus still hangs in the air.

Notwithstanding such unresolved issues, much in Deudney's analysis can be borrowed from, explored and built on for our purposes. For example, his Republican Security Theory may be repurposed to imagine scenarios where nuclear disarmament is realised. This can be done in relation to Deudney's concept of 'bounding power', a twofold idea referring to the massive advances in the power of military technology, which has increased beyond what was previously imaginable, and the need to put boundaries and controls on it via agreed restraints.[189] The value of this approach for the problems posed by nuclear disarmament is that it addresses interconnected security problems at the domestic and international levels. In this sense, it presents a fusion of the second and third images of Waltz, dealing with the causes of international conflict that exist at the unit (national) and system (international) level. Deudney thus sees the evolution of negarchy as involving 'the establishment of increasingly cobinding mutual restraints on armed force', which in the case of nuclear weapons might practically involve gradual arms control and disarmament measures.[190]

One meaning of, or scenario for, nuclear disarmament (which we may now provisionally propose following our discussion of Deudney's work) could therefore entail a transition from nuclear monarchies to non-nuclear democracies, concurrent with a transformed global security order where both interstate anarchy and hierarchical world government is eschewed in favour of a negarchic republican federation. Such a scenario clearly raises several big questions, not least of which is how we may differentiate between the changes required for nuclear disarmament in and between each NWS given their various political and strategic positions and processes, and the respective political forces that may be required to achieve such change. In order to begin our exploration of these questions, the next section will focus on the concept of *institutional democratisation*, which will help us both bring together and build on the insights gathered so far and further refine our understanding of what nuclear disarmament might mean in practice for NWS.

## 2.7 The domestic politics model and institutional democratisation

Having reviewed the work of several prominent realist authors and their critics, we may now both identify the main gaps in the literature and outline ideas for improving our understanding of the causes and consequences of NWS nuclear disarmament. In doing so, our inquiry will come to focus on the concept of *institutional democratisation*. Before outlining precisely what this idea entails, it is necessary to briefly summarise our findings from the preceding discussion. First, despite several authors—including from within the realist tradition—noting the domestic political impacts of nuclear possession, including on such crucial areas as liberty and democracy, their main focus when debating this subject remains on the international level and the ways in which anarchy is moderated by the balance of power and nuclear weapons.

Mainstream and realist international relations scholarship thus principally understands the advent of nuclear weapons as a 'revolutionary' development within the anarchic system, with great significance for national and international security. This is because the defence–offence equation was changed so that the great powers moved to deterrence-based relationships. According to most of, if not all, the authors reviewed above, the period of unprecedented great power peace and stability that existed during the Cold War may, albeit to an extent that cannot be precisely specified or proven, be significantly attributed to the impact of this nuclear revolution. Furthermore, for many of these authors, great power nuclear disarmament is thus both clearly undesirable, given the risks of instability and conflict it poses, but also likely unrealisable given the political and technical obstacles it faces.

A related question, which our reading of the realist literature raises, is whether NWS nuclear disarmament can occur within anarchy, for example, as a result of a two-track process of domestic political reform and international cooperation to mitigate fear and uncertainty, or whether it requires anarchy to be replaced by some sort of hierarchy, for example, a world government. Deudney's analysis complicates this question by adding a third structure—negarchy—to the mix. Would one of these three structures be better suited to nuclear disarmament than the other? In response to these questions, we may optimistically, but still cautiously, suggest at this early stage of our inquiry that nuclear disarmament could imaginably be possible within each structure and that negarchy, whilst under-specified as a means of supporting disarmament processes, is particularly promising.

Furthermore, as we have seen, of the three structures, there is much more informed debate about the consequences of anarchy and the possibility of mitigating its effects in order to achieve nuclear disarmament. But as with the ambivalent response of some scholars to the nuclear revolution itself, our conclusion regarding the possibility of great power nuclear disarmament under anarchy, as one of—if not the—biggest possible within-system changes, must remain hopeful yet inconclusive. Whilst this is a frustrating outcome, we should remind ourselves of the inherently speculative nature of our subject matter and aim to provide clearer responses to these questions after having conducted our review of nuclear politics in and between NWS in Part II of this study. Moreover, we may still constructively propose at this point that *institutional democratisation* provides important problem-solving tools within each structure, not least because of the ever-present need to hold powerful actors accountable for their actions and ensure that political processes enjoy popular legitimacy.

Leaving aside structural questions for now, several of the authors we reviewed have noted that nuclear weapons decision-making is—at least in part—driven by internal politics as well as external factors. Moreover, as Deudney makes clear, nuclear weapons have important domestic political consequences for the states that possess them. If we are therefore to imagine how nuclear disarmament might come about, it is incumbent upon us to look at both the internal *and* the external political arenas when discussing the meaning and significance of nuclear possession. Additionally, many of the authors reviewed above either do not fully outline what they imagine the causes and consequences of nuclear disarmament to be, nor provide well-rounded considerations of the goals of, and conditions and indicators for, disarmament.

Where authors do engage meaningfully with disarmament concepts—such as Glaser and, to an extent Craig and Deudney—there are gaps in their analysis, particularly at the domestic level. It is possible to explain the different reasons why this is the case for each author. Some, such as Mearsheimer and Waltz, are strongly supportive of the nuclear revolution, for example, because it benefits the US-led global order—which they generally see as liberal and benign, and because they believe nuclear possession contributed much to the maintenance of peace and stability during the Cold War. Other authors, such as Jervis, Glaser and Deudney, are much more ambivalent about the nuclear revolution—particularly given the dangers of the second nuclear age. And whilst some, such as Craig, Schelling and Herz, see nuclear disarmament as potentially beneficial, they also feel that its requirements—which may include: domestic political transformations; the development of effective conventional deterrence; robust monitoring and verification; and world government—present very steep, if not insurmountable, hurdles. Given the rejection of or scepticism regarding nuclear disarmament, mainstream international relations scholarship—and realist works in particular—thus understandably spends little or no time considering the domestic sources of political change that might facilitate nuclear disarmament nationally and internationally. This absence is ultimately a symptom of structural realism's pessimistic theory of political change, which focuses—albeit in different ways according to the author—on anarchy as the main factor determining a state's international behaviour.

One of the other main reasons why nuclear politics tend to be treated in this limited fashion is that the subject, in both mainstream and more critical security discourses, is very often viewed in terms of the material and military qualities of the weapons themselves. However, as scholars such as Jervis appreciate, of equal if not greater importance are the ideational and political implications of the bomb—which are much more difficult to measure, especially given the secrecy surrounding nuclear arsenals. The next step in our study must therefore be to find a means of investigating in more depth the domestic causes and consequences of nuclear *possession* in and between NWS in order to then understand the domestic causes and consequences of nuclear *disarmament* in and between NWS. First of all, this requires us to analyse the internal nuclear politics of each NWS to flesh out what Deudney's 'diagnosis of the misfit between the state system and nuclear explosives' means in practice for each NWS.[191] By undertaking this analysis we will then be much better placed to: i) appreciate the political obstacles to and opportunities for nuclear disarmament in and between the NWS, ii) consider what domestic and international political *changes* are required to advance and realise nuclear disarmament, including what domestic and international political *forces* are needed to drive such changes forward.

In order to consider the requirements of NWS taking unilateral, bilateral and multilateral nuclear disarmament action, in addition to looking at the circumstances of each NWS individually, this also means assessing the pertinence of different political solutions. For example, if the elimination of nuclear arsenals is to collectively occur for the NWS, will each NWS ultimately have to become a non-nuclear democracy within a global republican federation, or are there less extreme but still effective options? It is reasonable to suggest that the answer to this question will only become clear at some point after the disarmament process has begun and will likely depend on questions such as the level of irreversibility required for each NWS, the answer to which will inform what this process comes to mean politically for each NWS.

Returning to the question of how we can move beyond our previous discussion of mainstream international relations approaches to nuclear possession and disarmament, it is important to recognise that we will also need to revisit the issue of how nuclear disarmament has been justified, as previously outlined in Chapter 1. By reviewing the powerful objections to global nuclear disarmament relating to the stability of a disarmed world, and the supposed security dilemmas that this would generate, we have gone part of the way to accomplishing this task. This is because we can identify useful ideas and information from the literature that opposes, is critical or ambivalent towards the idea of nuclear disarmament—which we can adapt and borrow from—in addition to familiarising ourselves with key arguments we need to provide rebuttals to. The next step in constructing a new approach to nuclear disarmament involves drawing on ideas and analysis from those sympathetic to or supportive of nuclear abolition, particularly concerning the domestic and international causes and consequences of nuclear possession and disarmament and the way in which these two levels interact.

### 2.7.1  Relating institutional democratisation to nuclear disarmament

In conducting this initial survey of the pro-disarmament literature, we shall focus on the concept of *institutional democratisation*. This is a useful concept to deploy here because it both provides a potential solution to the absences and gaps we found in our previous literature review and because it captures the spirit of much of what the *disarmament first* literature and social movements (reviewed in Chapter 1) are based on and trying to accomplish. Before considering what *institutional democratisation* might mean in practice as a driver of disarmament and how it has been previously deployed, it is useful to define the term, building on ideas from the wider political and social science literature.

As David Held, William Hudson and Seva Gunitsky note, democracy is a complex concept, the definition of which—government or rule by the people—raises many questions.[192] Having reviewed several indexes, Gunitsky therefore observes that we should be 'extremely careful' when using different measures of democracy, given the highly contested and subjective nature of the topic. He recommends that we must also make clear the 'inherent biases' in the measure chosen to identify 'what causes democracy' and justify our choice 'in relation to what is actually being examined'.[193] As for Hudson, having conducted a review of four different models for democracy, he concludes that, although 'part of the meaning of democracy is a continued discussion of the meaning of democracy', certain 'common elements' may be discerned, namely, 'popular rule, equality and liberty'.[194]

The International Institute for Democracy and Electoral Assistance focuses on the two principles of popular control and equality, further breaking them down into five 'democratic attributes', namely: 'representative government, fundamental rights, checks on government, impartial administration and participatory engagement'.[195] Held, meanwhile, proposes that if democracy is to 'flourish' it needs to be reimagined as a 'double-sided phenomenon: concerned on the one hand, with the reform of state power and, on the other hand, with the restructuring of civil society'. This is necessary to ensure that the 'active citizen' can once more 'return to the centre of public life' and participate meaningfully in communal decision-making.[196]

Political institutions, meanwhile, have also been defined in different ways. For example, Konrad von Moltke sees them as 'social conventions or "rules of the game"', Marie Gottschalk as 'those formal organizations and procedures that determine "who gets what when and how" for a society', whilst Keohane describes them as 'a general pattern or categorization of activity' or a 'particular human-constructed arrangement, formally or informally organized'.[197] Combining these two terms together, we may propose that *institutional democratisation*—for the purposes of this study—refers to the process of making the institutions of each NWS that involve and are related to nuclear weapons decision-making, subject to democratic participation, oversight and control. Each NWS has a responsibility to ensure that they implement appropriate processes of democracy, transparency and accountability in order to enable the dismantling of their nuclear weapon systems, as required by the NPT. Such democratisation and disarmament measures are also required to realise universal human rights, including popular rule, equality and liberty domestically, and common and sustainable security globally.[198]

The importance of democratisation to moderating a state's actions has been outlined by Joshi, Maloy and Peterson in an article titled 'Popular vs. Elite Democratic Structures and International Peace', in which they distinguish between two explanations used by 'structural theories' to account for why democratic regimes tend towards peaceful behaviour. The 'popular logic' focuses on how democratic institutions expose 'policymaking elites to pressure from ordinary citizens'. The 'elite logic', meanwhile, focuses on 'pressure from other governmental agencies'. In order to isolate and test the popular logic empirically, the authors developed the Institutional Democracy Index, which they describe as 'a novel measure of institutionalized popular influence'.[199]

These authors find that there are 'significant differences within the family of democratic regimes' so that 'more popular democracies are less war-like with respect to all other regimes, not just other democracies' and argue that their methodology therefore allows them to capture the 'variance among democratic regimes in their structures of inclusion', which includes 'formal rules pertaining to voter access, electoral formulae, and cameral structures'. Such variance is significant as it allows us to observe 'crucial differences between the conflict propensities of more popular and more elite types of democracy'.[200] For example, France, the UK and US are identified as being amongst 'the more elite democracies' and the 'most powerful states militarily'.[201]

However, the authors do not consider the linkages between these two issues, for example, the degree to which the relative power and influence of the military—and militarism—impacts upon democratic processes, and the ability of elites to exercise political control. Notwithstanding this, the conclusion of the article is useful for studies such as mine, which consider what necessary domestic political reforms might be required in support of nuclear disarmament, by highlighting the need to consider whether 'more popular democratic institutions promote sustainable peace' as opposed to 'the more elite institutional options'.[202]

In a similar vein, Una Becker, Harald Müller and Simone Wisotzki have explored whether there is 'a particularly democratic way of dealing with nuclear arms control?' Drawing on democratic peace theory, their study argues that 'democracies should indeed develop a preference for arms control'. Importantly, the case studies they explore—which consist of six Western democracies, including France, the UK and US, reveal a 'considerable variance in their nuclear arms control policies'. They explain this variance via a social constructivist approach, which entails investigating these countries' 'respective roles, identities, and images of the Kantian "unjust enemy"'.[203] Significantly, for the purposes of this study, the authors find that 'the more explicitly states pronounce an "undemocratic enemy," the more prominent they consider nuclear weapons as a means for maintaining order, and the less active they are in nuclear arms control'.[204]

Moreover, according to these authors, the role France, the UK and US have historically occupied as global powers is particularly significant, because they see this as a 'responsibility' conferring 'special rights', including nuclear possession.[205] The variance in defence and foreign policy across democracies can also partly be explained by the different internal 'policy processes' these nations follow.[206] Whilst Becker, Müller and Wisotzki state that it is beyond the scope of their study to investigate the domestic politics of each of their case studies, their approach aligns with my central claim that *institutional democratisation* will enable NWS to make progress towards and realise nuclear disarmament action. Further support for this claim is provided by Isabella Alcañiz, who argues that the transition of states to democracy increases their incentives 'to commit to multilateral security agreements', which may be observed 'in the rate at which new democracies ratify international treaties of arms control, nuclear nonproliferation, and disarmament'.[207]

Robert Johansen has developed complementary ideas in his work, arguing that states 'need to transform National Security into Democratic Security'. This is because traditional concepts of security (i.e. related to realism) discourage democracy at the international, domestic and individual level. By contrast, democratic security 'begins with the assumption' that all are 'entitled' to collective and popular participation in 'those decisions that profoundly affect their lives'. In addition, democratic security must be for 'all people' and not for 'abstractions like the state' or 'particular governments, governing elites, and their immediate supporters', because it entails the protection of 'human dignity and widely agreed-upon human rights'.[208]

Notably, both Johansen and Held propose a cosmopolitan approach to democracy. For example, the former argues that global interdependence means that 'ultimately each of the world's governments must be held accountable to all people affected by its major decisions, whether they live within that government's borders or not', whilst the latter avers that 'national democracies require international democracy if they are to be sustained and developed in the contemporary era'.[209] Overall, Democratic Security as a concept may be seen as compatible with Deudney's Republican Security Theory, not least because Johansen discusses how traditional approaches to security legitimise military institutions that are 'hierarchical and authoritarian' giving 'a tiny, nuclear-armed set of governing elites an undemocratic, literal power over the future of civilization and life itself'.[210]

Given the discussion of *institutional democratisation* provided above, this concept can be seen as drawing on and developing the second of Scott Sagan's three models regarding why states build nuclear weapons, namely, the 'domestic politics model', which focuses on the 'domestic actors who encourage or discourage governments from pursuing the bomb'.[211] *Institutional democratisation* develops the domestic politics model in three important ways. First, *institutional democratisation*, as I use it in this study, explores the politics of nuclear possession and disarmament, whilst Sagan focuses on why states build the bomb. Secondly, Sagan's study does not consider the impact that nuclear possession has had over time on domestic politics within each NWS—which may be described as *nuclear development*—nor the preferences and responses of the citizens of NWS to their nation's nuclear status. Thirdly, Sagan observes that 'there is no well-developed domestic political theory of nuclear weapons proliferation' that can identify the conditions enabling coalitions of actors to be formed which can 'become powerful enough to produce their preferred outcomes' concerning a state's nuclear choices.[212] Through discussing the meaning of *institutional democratisation*, Part II of this study seeks to remedy this situation by specifying how domestic nuclear politics works in each NWS in order to consider what internal changes are required for progress on disarmament to be made.

### 2.7.2 Disarmament and democracy or nuclear possession and guardianship?

Robert Dahl provides a more explicitly democratic approach to the problem of nuclear possession in his *Controlling Nuclear Weapons: Democracy versus Guardianship*. In this work, which focuses on US nuclear politics, Dahl pursues a similar line of argument to Deudney, stating that because 'these weapons have largely escaped the control of the democratic process', US citizens have 'perhaps unwittingly adopted the Principle of Guardianship'.[213] The idea of guardianship refers to the notion that only 'a minority' of 'well qualified' people within a state are capable of making 'collective decisions'. Owing to the supposed complexity of nuclear weapons, Dahl proposes that US citizens have 'turned over' decisions in this area to 'a small group of people' and makes the blunt point that 'it is very far from clear how, if at all, we could recapture a control that in fact we have never had'.[214] Whilst Dahl provides an important investigation of how to solve the problems of nuclear possession and guardianship through democratic means, which include the development of a citizenry which has 'sufficient *competence*' to exercise 'sufficient *control*' regarding nuclear decisions, he does not link his study to the challenge of nuclear disarmament.[215]

However, Dahl's terminology is useful for this study and may be adapted to broadly distinguish between two different models of nuclear weapons decision-making. The first, which may be referred to as the *guardianship* model, can be said to refer to the idea that NWS decision-making elites alone are sufficiently capable of advancing nuclear arms control and disarmament and that, generally speaking, the citizenry are not capable of playing an active or significant role in this endeavour. As shall be discussed further in Part II of this study, the *guardianship* model of nuclear arms control and disarmament thus involves a series of incremental, step-by-step measures which NWS elites generally agree are necessary and sufficient, at the international level, to make progress on their NPT obligations.

The second, alternative model is *institutional democratisation*, which, as we've seen, refers to the idea that nuclear weapons decision-making should be controlled, directly or indirectly, by the citizens of NWS. Whilst there is some overlap between the two models, so that agreements and initiatives involving top-level decision-makers that contribute to military restraint and strategic stability are of vital importance, the *guardianship* model is, by its nature and location in the establishment, inherently limited in terms of the progress it can make on nuclear disarmament. It is therefore proposed that *institutional democratisation* is necessary to drive forward and enact the far-reaching political changes required if NWS are to eliminate their nuclear weapons.

Despite Dahl not discussing disarmament in any depth, the idea of using *institutional democratisation* to explain and understand nuclear disarmament does exist in the literature, although this exact term is not deployed, as far as I am aware. In addition to Cortright and Väyrynen's work, relevant ideas that we may draw on can be found in the works of authors and scholars such as Kennette Benedict, Elaine Scarry and Scilla Elworthy.[216] National and international civil society groups, which these and other authors have been involved in, have, since the dawn of the nuclear age, highlighted the role that public opinion—particularly regarding opposition to the bomb—should play in decisions of war and peace. Identifying where national and global citizenries stand on nuclear issues is surely a vital concern for any study of nuclear weapons decision-making, as the main source of legitimacy for governments on this vitally important topic, and shall be fully explored in Part II.

Applying ideas associated with *institutional democratisation* to the domestic political sphere as a means of making progress on nuclear disarmament has been proposed by civil society groups such as the UK-based Oxford Research Group since the early 1980s, but has been neglected or forgotten as an idea since then by more mainstream discussions of the topic, despite some other protest and research groups continuing with related efforts. Such civil society work on disarmament is based on a belief that the secrecy surrounding the development and reproduction of nuclear weapons, along with the highly centralised decision-making structures regarding their use, is incompatible with and deeply corrosive to the spirit and functioning of democracy. According to this analysis, for nuclear disarmament to succeed, each NWS must move towards a setup whereby the domestic political conditions that allow nuclear weapons to flourish are no longer present, or have been dramatically reduced. Pro-disarmament thinking has also connected the question of nuclear disarmament to the wider nature of NWS's defence and foreign policy. Such authors from the critical security, leftist and anti-imperialist tradition—particularly in the Western NWS—place principles such as democracy, social justice and international law at the centre of their analyses and calls for an alternative approach.[217]

Three veteran US disarmament activists I interviewed thus argued that nuclear disarmament is not just about eliminating nuclear weapons themselves but, especially in the case of the US, requires major societal changes given the centrality of the nuclear weapons complex to the economy and military culture. Nuclear disarmament will therefore entail 'transforming domestic power structures' to 'redistribute wealth and power' as part of a democratic process of moving to 'environmentally sustainable economics'. For the US this will mean, they argue, 'popular social movements' creating a 'change in culture and ethics' to reduce the power of the 'military-industrial complex'.[218]

In order to apply ideas and approaches from pro-disarmament movements and make them relevant to today's environment, it is necessary to conduct a review of nuclear politics in and between the NWS to develop a theory of nuclear disarmament that is both rigorous and up to date. I shall explain further in Part II how this investigation shall proceed, suffice to say here that important considerations which need to be taken into account to illustrate the value of an approach based on *institutional democratisation* include: i) the relative strategic power of each NWS and the nature of their international relations, particularly with other NPS, ii) the regime type of each NWS, for example, more or less authoritarian or democratic, iii) the causes and consequences of nuclear possession and disarmament for each NWS.

As we have seen, authors working in the realist tradition are sceptical of the potential for democratisation alone to succeed in creating the conditions for international peace and security. For example, Niebuhr argues that anarchy will prevail until all the great powers became democratic.[219] Secondly, Waltz draws on the historical record to argue that, even if a state is or becomes democratic it will still be prone to conflict under anarchy. He also argues that it is unrealistic for international peace to rely on all democracies being of a similar nature. Several points can be made in response to these observations, the first being that it is not possible to make a definitive judgement about the future impacts of democratisation on the international system. Secondly, there needs to be some objective measurement of how democratic a state is in order to make useful assessments, principally because Waltz's case studies do not include modern democracies or those with more popular and less elitist democratic processes. Thirdly, Deudney's Republican Security Theory goes some way to potentially solving the problems posed by both anarchy and hierarchy by proposing a third way, i.e. negarchy. However, Deudney does not himself propose a theory of nuclear disarmament or one sympathetic to it and, despite emphasising the need for democratic legitimacy, does not develop a theory of political change to explain how his proposals may be realised.

Having briefly considered what *institutional democratisation* pursuant to nuclear disarmament might mean on a domestic level, based on past and present work from academic and civil society, it is appropriate to finish this section by outlining the work of international civil society and global governance with relevance to nuclear possession and disarmament. This is also important to consider in order to identify potential international sources of political change. As noted above, in addition to nationally based civil society groups (primarily in France, the UK and US) engaging in pro-disarmament action—including education, protest and research, as reviewed in the unique historical work of Lawrence Wittner—a vibrant international peace and disarmament movement exists whose ideas and proposals need to be taken into account.[220]

Yet, as we've seen, realist thought often sidelines such considerations, instead claiming that the domestic nature of states, regimes, groups or individuals is irrelevant to nuclear decisions or outcomes.[221] Indeed, Mearsheimer posits that actors can be treated as 'black boxes or billiard balls' because the international system itself causes conflict.[222] Ole Holsti argues that 'realists' of all stripes (including Morgenthau in his earlier writings) have thus generally seen public opinion as 'largely irrelevant in the conduct of foreign affairs' because of its volatility and lack of coherent structure. Importantly, whereas the public has preferences which are 'allegedly driven by emotions and short-term considerations', realist thought has tended to present officials and leaders as rational and cool-headed.[223]

However, research conducted in several democracies, including in Europe, Japan and the US, has found that, where anti-nuclear weapons public opinion, protest and civil society activism exists, it has exerted an influence on the degree and type of action taken by governments. This corresponds to Thomas Risse-Kappen's finding that, whilst mass public opinion in liberal democracies rarely has a 'direct affect' on 'policy decisions or the implementation of specific policies', it does 'set broad and unspecified limits' to 'foreign policy choices'.[224] It has thus been argued, by various authors, that civil society action has been significantly responsible for increased cooperation on arms control, the creation of nuclear weapons free zones and moves to ban nuclear testing, the development of a taboo against the use of nuclear weapons and the decision by a number of states to exercise restraint and not seek to acquire the bomb.[225] Rather than being emotional and unstable therefore, Holsti posits a 'rational public' thesis, suggesting the public is a source of 'moderation and continuity rather than of instability and unpredictability'.[226]

## 2.8 Chapter summary

This chapter's review of the range of approaches from within and without realist thought that engage with the advent of the bomb and the nuclear revolution made several useful discoveries. For example, each of these approaches contained both descriptive and prescriptive elements, with authors basing their response to the nuclear revolution on what they see as most beneficial for national and international security and the avoidance of nuclear war. However, in doing so, US security was treated as paramount by many of these authors, with some, such as Mearsheimer, advocating a particularly nationalistic approach. Meanwhile, nuclear disarmament was generally either seen as an enterprise that is undesirable and unrealisable—or potentially beneficial, albeit with several strong qualifications and reservations attached. These barriers to disarmament, it was argued, derive from the significant political, technical and diplomatic difficulties faced at the domestic and international levels. Whilst none of the authors reviewed in depth thus explicitly advocated nuclear disarmament, significantly engaged with the pro-disarmament literature and its arguments, differentiated between its unilateral, bilateral and multilateral forms in their discussion or dealt substantively with the domestic politics of nuclear possession and disarmament, each still raises important issues that any theory of nuclear disarmament must engage with in order to provide a well-rounded analysis.

These issues include those both of a more abstract and theoretical as well as a more concrete and political nature, such as: i) whether nuclear disarmament within anarchy is possible or whether it can only happen within negarchy or hierarchy, ii) whether nuclear disarmament would necessarily lead to anarchy being replaced by another system and the type of domestic, regional and global political and security frameworks it necessitates, iii) how regional and international peace and security can best be managed before, during and after nuclear disarmament, iv) the principles on which political change supportive of nuclear disarmament should be based, for example, democracy, transparency and accountability, v) the extent to which nuclear disarmament requires domestic *and* international political change, including sustained interstate cooperation to mitigate fear and uncertainty, vi) the size and scale of the domestic and international political forces required to realise progressive nuclear disarmament action, vii) how national, international and universal political identities and institutions can be reconciled.

Having identified these problems and questions, our next logical move would be to propose an alternate approach or theory of nuclear disarmament that may be tested against empirical evidence, which in this instance would centre on nuclear politics in and between the NWS. Regarding the latter exercise, which we began to outline above, the fact remains that any investigation of the political causes and consequences of complete nuclear disarmament by one or more NWS would be necessarily speculative since it covers something that has and is not happening. We are thus faced with the methodological problem of how to 'test' any theory we may construct. For example, our preceding discussion leads us to make two important proposals. First, that the invention and development of nuclear weapons led to far-reaching political changes occurring at both the international *and* domestic levels, but the implications for the latter have not been properly recognised by mainstream international relations and political scholars nor fully explored historically, or in relation to the political institutions of the present day. Secondly, that in order for nuclear disarmament to occur, appropriate, concurrent and complementary political change needs to take place at both levels.

To this end, *institutional democratisation* is a concept which could both fill the gaps identified in our literature review and which provides a way forward by suggesting the type of legitimate and necessary political forces and processes that nuclear disarmament requires and will benefit from at both the domestic and international levels. Testing this theory would first mean investigating the causes and consequences of nuclear possession in and between the NWS to show how an analysis that takes the domestic impact of nuclear possession into account adds significant explanatory value in terms of each of these states' nuclear histories and present circumstances. Having done this, the second, more normative and speculative step, would be to investigate the strengths and weaknesses of *institutional democratisation* as a theory of political change that could support the goal of nuclear disarmament in and between the NWS. Part II provides this investigation through an analysis of five case studies, covering each of the NWS in order of them acquiring the bomb: the US, Russia, UK, France and China.

## Notes

1  Jervis, Robert (1986) The Nuclear Revolution and the Common Defense, *Political Science Quarterly*, 101(5), Reflections on Providing for 'The Common Defense', 689–690.
2  Sagan, Scott D. (1996) Why Do States Build Nuclear Weapons? Three Models in Search of a Bomb, *International Security*, 21(3), Winter, 57.
3  Mearsheimer, John (2006) Conversations in International Relations: Interview with John J. Mearsheimer (Part I), *International Relations*, 20(1), 114.
4  Mearsheimer, John (2001) *The Tragedy of Great Power Politics* (New York: Norton), 29.
5  Ibid., 2.
6  Ibid., 30–32.
7  Ibid., 35–36; Herz, John H. (1951) *Political Realism and Political Idealism: A Study in Theories and Realities* (Chicago, IL: University of Chicago Press).
8  Booth, Ken and Wheeler, Nicholas J. (2008) *The Security Dilemma: Fear, Cooperation and Trust in World Politics* (Basingstoke: Palgrave), 1–9.
9  Mearsheimer, *Tragedy of Great Power Politics*, 32.
10  Mearsheimer, Conversations in International Relations (Part I), 118–123.
11  Mearsheimer, *Tragedy of Great Power Politics*, 358–359.
12  Ibid.
13  Perry, William J. (2015) *My Journey at the Nuclear Brink* (Stanford, CA: Stanford University Press), 5.
14  Mearsheimer, Conversations in International Relations (Part I), 111.
15  Mearsheimer, *Tragedy of Great Power Politics*, 51; Mearsheimer, John (2006) Conversations in International Relations: Interview with John J. Mearsheimer (Part II), *International Relations*, 20(1), 233.

16 Mearsheimer, John (2006) Conversations in International Relations: Interview with John J. Mearsheimer (Part II), *International Relations*, Vol 20(1), 233.
17 Mearsheimer, *Tragedy of Great Power Politics*, 50.
18 Mearsheimer, John (1990) Why We Will Soon Miss the Cold War, www.theatlantic. com, August; *Tragedy of Great Power Politics*, 130.
19 Mearsheimer, Miss the Cold War.
20 Mearsheimer, *Tragedy of Great Power Politics*, 367.
21 Ibid., 363.
22 Mearsheimer, Conversations in International Relations (Part II), 237.
23 Mearsheimer, *Tragedy of Great Power Politics*, 130, 146.
24 Ibid., 392.
25 Mearsheimer, Miss the Cold War.
26 Ibid.; Mearsheimer, Conversations in International Relations (Part II), 240.
27 Ibid.; Mearsheimer, John (2006) Conversations in International Relations (Part II), 121.
28 Mearsheimer, John (2015) What Is America's Purpose? *The National Interest*, September/October, 35.
29 Herz, John H. (1959) *International Politics in the Atomic Age* (New York: Columbia University Press).
30 Ibid., v, 331–336.
31 Ibid., 197, 339.
32 Ibid., 335.
33 Mearsheimer, *Tragedy of Great Power Politics*, 17.
34 Mearsheimer, Miss the Cold War.
35 Hardin, Russell and Mearsheimer, John (1985) 'Introduction' in Symposium on Ethics and Deterrence, *Ethics*, 95(3), April, 422.
36 Mearsheimer, *Tragedy of Great Power Politics*, 367, 406.
37 Ibid., 406.
38 Ibid., 402.
39 Peceny, Mark, Beer, Caroline and Sanchez-Terry, Shannon (2002) Dictatorial Peace? *American Political Science Review*, 96, 15–26; Rosato, Sebastian (2003) The Flawed Logic of Democratic Peace Theory, *American Political Science Review*, 97(4), November.
40 Jervis, Robert (1978) Cooperation under the Security Dilemma, *World Politics*, 30(2), January, 167.
41 Jervis, Robert (1998) Realism in the Study of World Politics, *International Organization*, 52(4), Autumn, 986.
42 Jervis, Cooperation, 187.
43 Glaser, Charles (1997) The Security Dilemma Revisited, *World Politics*, 50(1), October, 189.
44 Ibid., 171.
45 Jervis, Cooperation, 214.
46 Ibid., 201.
47 Glaser, Security Dilemma Revisited, 174, 189, 192, 193.
48 Jervis, Robert (1979) Why Nuclear Superiority Doesn't Matter, *Political Science Quarterly*, 94(4), Winter, 618.
49 Jervis, Cooperation, 206.
50 Jervis, Robert (2009) Unipolarity: A Structural Perspective, *World Politics*, 61(1), January, 201.

51  Jervis, Robert (1988) The Political Effects of Nuclear Weapons: A Comment, *International Security*, 13(2), Fall, 81; Jervis, Nuclear Revolution and Common Defense, 694.
52  Jervis, Nuclear Revolution and Common Defense, 702.
53  Jervis, Cooperation, 214.
54  Jervis, Nuclear Revolution and Common Defense, 702.
55  Jervis, Robert (1989) *The Meaning of the Nuclear Revolution: Statecraft and the Prospect of Armageddon* (Ithaca, NY: Cornell), 257.
56  Jervis, Nuclear Revolution and Common Defense, 695.
57  Ibid., 701.
58  Jervis, *Meaning of the Nuclear Revolution*, 25–27.
59  Jervis, Robert (1999) Realism, Neoliberalism, and Cooperation Understanding the Debate, *International Security*, 24(1), Summer, 42.
60  Ibid., 45.
61  Ibid., 61.
62  Ibid., 45.
63  Jervis, Robert (1994) Hans Morgenthau, Realism, and the Scientific Study of International Politics, *Social Research*, 61(4), Winter, 872.
64  Jervis, Cooperation, 175–176.
65  Ibid.
66  Glaser, Charles (1998) The Flawed Case for Nuclear Disarmament, *Survival*, 40(1), Spring, 112.
67  Glaser, Charles (2007) The Instability of Small Numbers Revisited: Prospects for Disarmament and Nonproliferation, in Michael May, ed., *Rebuilding the NPT Consensus*, 16–17 October, Center for International Security and Cooperation (Stanford, CA: Stanford University), 222.
68  Glaser, Charles (1990) *Analyzing Strategic Nuclear Policy* (Princeton, NJ: Princeton University Press), 11, 181–183.
69  Glaser, Flawed Case, 113, 117; Glaser, Instability of Small Numbers, 218.
70  Schelling, Thomas (2009) A World without Nuclear Weapons?, *Daedalus*, 138(4), On the Global Nuclear Future, 1, Fall, 125.
71  Ibid., 127.
72  Ibid., 126.
73  Schelling, Thomas (1962) The Role of Deterrence in Total Disarmament, *Foreign Affairs*, 40(3), April, 401–402.
74  Ibid., 406.
75  Jervis, Political Effects, 83.
76  Glaser, Security Dilemma Revisited, 183.
77  Glaser, Flawed Case, 119.
78  Glaser, Instability of Small Numbers, 219.
79  Waltz, Kenneth (2010) *Theory of International Politics* (Long Grove; Waveland), 89.
80  Waltz, Kenneth (1993) The Emerging Structure of International Politics, *International Security*, 18(2), Autumn, 49; Mearsheimer, *Tragedy of Great Power Politics*, 19.
81  Waltz, *Theory of International Politics*, 91.
82  Ibid., 100.
83  Ibid., 95, 97.
84  Waltz, Kenneth (2008) *Realism and International Politics* (Abingdon: Routledge), 41.

85 Schweller, Randall (1996) Neorealism's Status-Quo Bias: What Security Dilemma? *Security Studies*, 5(3), 90–93.
86 Ibid.
87 Waltz, *Theory of International Politics*, 91.
88 Waltz, Kenneth (2000) Structural Realism after the Cold War, *International Security*, 25(1), Summer, 23, 28.
89 Schweller, Neorealism's Status-Quo Bias, 103.
90 Waltz, *Theory of International Politics*, 161–163.
91 Waltz, Emerging Structure, 44.
92 Waltz, Structural Realism, 28, 37.
93 Waltz, *Theory of International Politics*, 196.
94 Waltz, Structural Realism, 26.
95 Waltz, Emerging Structure, 77.
96 Waltz, *Realism and International Politics*, xii
97 Waltz, Structural, 10.
98 Jervis, Robert (2014) Interview with Robert Jervis, *International Relations*, 28(4), 500.
99 Waltz, Structural Realism, 13.
100 Waltz, *Realism and International Politics*, xii.
101 Waltz, Kenneth (1990) Nuclear Myths and Political Realities, *American Political Science Review*, 84(3), September, 741.
102 Craig, Campbell (2003) *Glimmer of a New Leviathan: Total War in the Realism of Niebuhr, Morgenthau and Waltz* (New York: Columbia University Press), 169–170.
103 Waltz, Structural Realism, 5.
104 Lieber, Keir A. and Press, Daryl G. (2006) The Rise of US Nuclear Primacy, *Foreign Affairs*, 85(2), March/April, 42.
105 Waltz, *Theory of International Politics*, 187.
106 Waltz, Kenneth (1997) Thoughts about Virtual Nuclear Arsenals, *Washington Quarterly*, 20(3), 154.
107 Sagan, Scott and Waltz, Kenneth (2013) *The Spread of Nuclear Weapons: An Enduring Debate* (New York: W. W. Norton), 220–224.
108 Ibid., 36.
109 Ibid., 221.
110 Paul, T. V. (2000) *Power versus Prudence: Why Nations Forgo Nuclear Weapons* (Quebec City: McGill-Queen's University Press), 8.
111 Rublee, Maria Rost (2009) *Nonproliferation Norms: Why States Choose Nuclear Restraint* (Athens, GA: University of Georgia), 10.
112 Keck, Zachary (2012) Kenneth Waltz on 'Why Iran Should Get the Bomb', https://thediplomat.com, 8 July.
113 Waltz, Kenneth (1995) *Policy Paper 15: Peace, Stability, and Nuclear Weapons*, Institute on Global Conflict and Cooperation, University of California, 13 August, 8, 14.
114 Sagan and Waltz, *Spread of Nuclear Weapons*, 10.
115 Ibid., 37.
116 Perry, *My Journey*, 81; Blackham, Jeremy (2013) *Deterrence Is Not Just about Nuclear Weapons: Time for Serious Strategic Thought* (London: UK National Defence Association, August).

117  Sagan and Waltz *Spread of Nuclear Weapons*, 77.
118  Waltz, Virtual Nuclear Arsenals, 160.
119  Waltz, Emerging Structure, 73.
120  Sagan and Waltz, *Spread of Nuclear Weapons*, 223.
121  Waltz, Kenneth and Fearon, James (2012), A Conversation with Kenneth Waltz, *Annual Review of Political Science*, 15, 9.
122  Keohane, Robert (2009) Big Questions in the Study of World Politics, in Goodin, Robert E., ed., *The Oxford Handbook of Political Science* (Oxford: Oxford University Press), 770,
123  Keohane, Robert (1984) *After Hegemony: Cooperation and Discord in the World Political Economy* (Princeton, NJ: Princeton University Press), 26.
124  Keohane, Robert (1983) Theory of World Politics: Structural Realism and Beyond, in Finiter, Ada W., ed., *Political Science: The State of the Discipline*, (Washington, DC: American Political Science Association), 532.
125  Keohane, *After Hegemony*, 51.
126  Axelrod, Robert and Keohane, Robert (1985) Achieving Cooperation under Anarchy: Strategies and Institutions, *World Politics*, 38(1), October, 236.
127  Keohane, Big Questions, 774.
128  Keohane, *After Hegemony*, 13.
129  Axelrod and Keohane, Achieving Cooperation, 226.
130  Ibid., 250.
131  Ibid., 241.
132  Keohane, Robert and Milner, Helen V. eds. (1996) *Internationalization and Domestic Politics* (Cambridge: Cambridge University Press), 4.
133  Ibid., 6.
134  Ibid., 7.
135  Cortright, David and Väyrynen, Raimo (2009) *Towards Nuclear Zero* (London: Routledge), 93; Walker, William (2012) *A Perpetual Menace: Nuclear Weapons and International Order* (London: Routledge), 163; Müller, Harald (2009) The Importance of Framework Conditions, in Perkovich, George and Acton, James M. eds., *Abolishing Nuclear Weapons: A Debate* (Washington, DC: Carnegie Endowment for International Peace).
136  Cortright and Väyrynen, *Towards Nuclear Zero*, 160–161.
137  Perkovich, George and Acton, James M., eds. (2009) *Abolishing Nuclear Weapons: A Debate* (Washington, DC: Carnegie Endowment for International Peace), 13.
138  Ibid., 23.
139  Wheeler, Nicholas J. (2009) *Nuclear Abolition: Trust-Building's Greatest Challenge?* (Canberra: ICNND).
140  Ibid., 3; Wheeler, Nicholas J. (2010) Trust-Building and the Nuclear Future, *Networking Symposium Report, 7th–8th September* (Aberwystwyth: Aberwystwyth University); Wheeler (2011) Nuclear Rivalries: Prospects for Cooperation and Trust-Building, *Networking Symposium Report, 14th–15th June*, (Aberwystwyth: Aberwystwyth University); Wheeler (2012) Trust-Building in International Relations, *Peace Prints: South Asian Journal of Peacebuilding*, 4(2), Winter.
141  Wheeler, *Nuclear Abolition*, 5.
142  Ibid., 2.

143 Freedman, Lawrence (2009) Nuclear Disarmament: From a Popular Movement to an Elite Project, and Back Again? in Perkovich and Acton, *Abolishing Nuclear Weapons*, 143.
144 Ibid., 144.
145 Nebel, Jacob (2012) The Nuclear Disarmament Movement: Politics, Potential, and Strategy, *Journal of Peace Education*, 9, 3, 231.
146 Ibid., 232.
147 Wheeler, *Nuclear Abolition*, 19–20.
148 Ibid.
149 Ibid.
150 Nuclear Threat Initiative (2017) *Treaty on the Prohibition of Nuclear Weapons*, www.nti.org, 6 October.
151 Kreisler, Harry (2004) *Globalization and 9/11*, http://globetrotter.berkeley.edu, 9 March, 5.
152 Keohane and Milner, *Internationalization*, 6.
153 Craig, Campbell (2003) *Glimmer of a New Leviathan: Total War in the Realism of Niebuhr, Morgenthau and Waltz* (New York: Columbia University Press), ix–xi.
154 Ibid: xi.
155 Craig, Campbell and Radchenko, Sergey (2017) MAD, Not Marx: Khrushchev and the Nuclear Revolution, *Journal of Strategic Studies*, 41(1–2), 23.
156 Craig, *Glimmer*, x.
157 Ibid., 129, 172.
158 Ibid., 172.
159 Ritchie, Nick (2012) *A Nuclear Weapons-Free World? Britain, Trident and the Challenges Ahead* (Basingstoke: Palgrave Macmillan), 28.
160 Craig, *Glimmer*, 171; Craig, Campbell (2015) *Weberian World Government in the Nuclear Age*, http://wgresearch.org, 12 June.
161 Craig, *Weberian World Government*.
162 Craig, Campbell (2009) American Power Preponderance and the Nuclear Revolution, *Review of International Studies*, 35(1), 27, 35.
163 Craig, *Glimmer*, 49, 66.
164 Ibid., 164.
165 Craig, *Weberian World Government*.
166 Ibid.
167 Craig, Campbell and Radchenko, Sergey (2008) *The Atomic Bomb and the Origins of the Cold War* (New Haven, CT: Yale University Press), 169.
168 Ibid.
169 Ibid., 170.
170 Craig, Campbell (2008) The Resurgent Idea of World Government, *Ethics and International Affairs*, 22(2), Summer, 7 July.
171 Deudney, Daniel H. (2011) Unipolarity and Nuclear Weapons, in Ikenberry, G., Mastanduno, M. and Wohlforth, W., eds., *International Relations Theory and the Consequences of Unipolarity* (Cambridge: Cambridge University Press), 282.
172 Ibid., 283.
173 Ibid., 284.
174 Ibid., 287.
175 Deudney, Daniel H. (2016) An Interview with Daniel Deudney, http://wgresearch.org.

176   Deudney, Daniel H. (2007) *Bounding Power: Republican Security Theory from the Polis to the Global Village* (Princeton, NJ: Princeton University Press), 48.
177   Ibid.
178   Ibid., 270.
179   Ibid., 255.
180   Ibid., 7.
181   Ibid., 8.
182   Ibid., 47.
183   Ibid., 50.
184   Ibid., 258.
185   Ibid., 246.
186   Ibid., 248.
187   Ibid., 259.
188   Ibid., 249.
189   Ibid., 27–28.
190   Ibid., 50.
191   Ibid., 250.
192   Held, David (1992) Democracy: From City-States to a Cosmopolitan Order, in Prospects for Democracy, Held, David, ed., *Political Studies*, 40(10); Hudson, William E. (2013) *American Democracy in Peril: Eight Challenges to America's Future*, 7th Edition (London: Sage), 1.
193   Gunitsky, Seva (2015) How Do you Measure 'Democracy'? www.washingtonpost.com, 23rd June.
194   Hudson, *American Democracy in Peril*, 20–21.
195   International IDEA (2017), The Global State of Democracy Indices, www.idea.int.
196   Held, Democracy: From City-States, 20.
197   von Moltke, Konrad (2001) Whither MEAs? The Role of International Environmental Management in the Trade and Environment Agenda, www.iisd.org, 11 July; Gottschalk, Marie (2000) *The Shadow Welfare State: Labor, Business and the Politics of Health Care in the United States* (Ithaca, NY: ILR/Cornell University Press), 5; Keohane, Robert (1998) International Institutions: Can Interdependence Work?, *Foreign Policy*, Spring, 383.
198   United Nations (1948), *Universal Declaration of Human Rights*, www.un.org.
199   Joshi, Devin K., Maloy, J. S. and Peterson, Timothy M. (2015) Popular vs. Elite Democratic Structures and International Peace, *Journal of Peace Research*, 52(4), 463.
200   Ibid.
201   Ibid., 470.
202   Ibid., 475.
203   Becker, Una, Müller, Harald and Wisotzki, Simone (2008) Democracy and Nuclear Arms Control—Destiny or Ambiguity?, *Security Studies*, 17(4), 810.
204   Ibid., 847.
205   Ibid., 825, 828, 833.
206   Ibid., 853.
207   Alcañiz, Isabella (2012) Democratization and Multilateral Security, *World Politics*, 64(2), April, 306.
208   Johansen, Robert C. (1991) Real Security Is Democratic Security, *Alternatives: Global, Local, Political*, 16(2), 210–212; Johansen, Robert C. (1992)

Military Policies and the State System as Impediments to Democracy, Prospects for Democracy, Held, David ed., *Political Studies*, 40, 113.

209 Johansen, Real Security, 212; Held, Democracy: From City-States, 11.

210 Johansen, Military Policies, 212.

211 Sagan, Scott D. (1996) Why Do States Build, 63.

212 Ibid., 64.

213 Dahl, Robert A. (1985) *Controlling Nuclear Weapons: Democracy Versus Guardianship* (Syracuse, NY: Syracuse University Press), 3–7.

214 Ibid., 7.

215 Ibid., 72.

216 Benedict, Kennette (2016) Add Democracy to Nuclear Policy, in Tom Z. Collina and Geoff Wilson, eds., *10 Big Nuclear Ideas for the Next President* (San Francisco, CA: Ploughshares Fund); Scarry, Elaine (2014) *Thermonuclear Monarchy: Choosing between Democracy and Doom* (New York: Norton); McLean, Scilla, ed. (1986) *How Nuclear Weapons Decisions Are Made* (Basingstoke: Macmillan).

217 Gittings, John and Davis, Ian, eds. (1997) *Rethinking Defence and Foreign Policy* (Nottingham: Spokesman); Monbiot, George (2006) Only Paranoia can Justify the World's Second Biggest Military Budget, www.theguardian.com, 28 November; Falk, Richard (2014) *(Re)Imagining Humane Global Governance* (London: Routledge).

218 Interviewees: A; B; C.

219 Craig, *Glimmer*, 52.

220 Wittner, Lawrence (2009) *Confronting the Bomb: A Short History of the World Nuclear Disarmament Movement* (Stanford, CA: Stanford University Press).

221 Solingen, Etel (2007) *Nuclear Logics: Contrasting Paths in East Asia and the Middle East* (Princeton, NJ: Princeton University Press), 11.

222 Mearsheimer, *Tragedy of Great Power Politics*, 10–11.

223 Holsti, Ole R. (1996) Public Opinion on Human Rights in American Foreign Policy, http://americandiplomacy.web.unc.edu, September.

224 Risse-Kappen, Thomas (1991) Public Opinion, Domestic Structure, and Foreign Policy in Liberal Democracies, *World Politics*, 43(4), July, 510.

225 Knopf, Jeffrey W. (1998) *Domestic Society and International Cooperation: The Impact of Protest on US Arms Control Policy* (Cambridge: Cambridge University Press); Wittner, Lawrence (2009) *Confronting the Bomb: A Short History of the World Nuclear Disarmament Movement* (Stanford, CA: Stanford University Press); Tannenwald, Nina (2008) *The Nuclear Taboo: The United States and the Non-Use of Nuclear Weapons since 1945* (Cambridge: Cambridge University Press); Rublee, Maria Rost (2009) *Nonproliferation Norms: Why States Choose Nuclear Restraint* (Athens, GA: University of Georgia).

226 Holsti, Ole R. (1992) Public Opinion and Foreign Policy: Challenges to the Almond Lippmann Consensus, *International Studies Quarterly*, 36(4), December, 447.

# Part II

# Obstacles to and opportunities for nuclear disarmament in and between the nuclear weapon states

In Part I many of the key problems associated with nuclear disarmament were introduced and its meaning and significance, and the different ways it has been conceptualised, discussed. One key finding was that mainstream and realist works on the causes and consequences of nuclear possession and disarmament mainly focus on the military-security challenges presented by the anarchic international system, with abolition largely seen as an undesirable and distant, if not impossible, prospect. Due to the limitations of these analyses, the concept of *institutional democratisation* was developed as a means of rebalancing the study of nuclear politics towards the domestic level as a precursor to exploring how domestic and international nuclear politics interact to prevent NWS nuclear disarmament taking place. Moreover, *institutional democratisation* also provides a coherent theory of political change for nuclear abolition in order to build on, refine and update existing ideas and proposals from the critical and pro-disarmament literature.

In order to provide some means of testing the explanatory power of *institutional democratisation*, we shall, in Chapters 3 to 7, engage in a series of in-depth analyses of the nuclear weapons systems of each of the five NWS in order of their acquiring the bomb: the US, Russia, UK, France and China. The point of doing this is to further show how the various existing theories and approaches outlined in Chapter 2 fail to capture the political dynamics that frustrate nuclear disarmament and how these might be overcome through adopting an approach based on *institutional democratisation*. Chapter 8 then summarises the relationships and interactions between NWS, contextualising this in terms of the wider sphere of global nuclear order, and outlines several ideas and proposals supportive of *institutional democratisation* on a national level.

In terms of the flow of the five NWS case studies presented below, each has broadly the same structure, beginning with a section that provides an overview of how mainstream and realist thought explain the politics of nuclear possession and disarmament for the NWS in question. This is followed by a section which provides an overview of how *institutional democratisation*—and ideas supportive of this concept—gives an alternative viewpoint on this issue. Both of these sections examine these approaches in relation to the historical record and contemporary scholarship with the goal of showing how *institutional democratisation* brings value to the existing literature in relation to the causes and consequences of NWS nuclear disarmament. In addition, both sections have a historical focus because they explore different perspectives on the origins and development of the bomb in each NWS, for example, in the context of WW2 and the Cold War. This is done to better understand the current meanings and value assigned to nuclear possession by elites, and other key actors in each NWS, such as the public, in order to then appreciate the politics of nuclear disarmament for each NWS.

In outlining the limitations to mainstream and realist explanations of nuclear possession and disarmament for each NWS, the point is not to throw out entirely the claims and ideas contained therein, for example, regarding the flaws of *cooperation with disarmament* and the *guardianship* approach to arms control and disarmament. Rather, our aim is to highlight and increase in emphasis the important role domestic political factors play in nuclear weapons decision-making as the precursor to discussing and evaluating the validity of *institutional democratisation* for a nuclear disarmament process applicable to the circumstances of each NWS. Having identified and compared different perspectives on the history of NWS nuclear possession, each chapter then includes a section that applies the theory of *institutional democratisation* to post-Cold War nuclear politics for each NWS. This is done, for example, by focusing on the domestic political obstacles to and opportunities for disarmament for each NWS today, in addition to reviewing the contemporary strategic challenges, and problems of cooperation relating to nuclear disarmament that the NWS face.

In terms of the size of these case studies, the chapters for Russia and the US are far lengthier than the chapters on the UK, France and China. There are several reasons for this, including the much greater size of the nuclear arsenals of Russia and the US, the fact that these two nations were the key participants in the Cold War nuclear standoff, and the availability and scale of the literature on nuclear issues for these two cases. Given the importance of Cold War politics to the development of nuclear weapons in the US and the then Soviet Union, and the impact this conflict had on the other NWS, I also include, for these two case studies, a discussion of competing explanations of the origins of the Cold War, based on what Andrew Kydd describes as the 'motivations and beliefs of the two sides', in order to clarify the differences between the various perspectives on nuclear politics.[1]

For example, I argue that, whilst the mainstream and realist approaches reviewed in Chapter 2 have diverse outlooks, generally speaking, they best correspond to what Kydd describes as 'post-revisionist' or 'traditionalist' perspectives on the meaning of the Cold War and its aftermath.[2] The former position traces the origins of this conflict to 'mistrust', so that the two superpowers were both security seekers, but their 'desire for security propelled them into conflict' and a highly tense nuclear standoff.[3] As for traditionalism, this is more commonly found in the non-scholarly literature, including from more hawkish US military and government sources. As Kydd notes, this position presents the Soviet Union as an 'inherently expansionist power, interested in exploiting cooperation, not reciprocating it, that is, untrustworthy', whilst the US is 'coded as a security seeker; trusting at first, and then increasingly fearful as time went by'.[4]

The specific claims and ideas of *institutional democratisation*, meanwhile, correspond better to revisionist and other critical perspectives on the Cold War and US global strategy. For example, revisionist accounts of the Cold War, as Kydd explains, argue that the Soviet Union was 'primarily defensively motivated while the capitalist West was the imperialistic and aggressive party'. 'Orwellianism', meanwhile, contends that both superpowers were expansionist so that the Cold War was 'nontragic'.[5] Notably, it is often the case that the critical nuclear literature, which is supportive of or compatible with *institutional democratisation*, adopts a revisionist or Orwellian approach to the Cold War.

In addition, for the US chapter, a section is included that highlights the idea of the NWS—and the US in particular—having dual nuclear disarmament responsibilities, both to eliminate their nuclear arsenals (and thus transition to FNWS status) and act in ways which support the creation of a NWFW. Regarding China, France, the UK and Russia's relative strategic power, these nations have relatively limited power projection capabilities, so they do not play a comparable role to the US in terms of other state's strategic thinking, including regarding nuclear weapons. Therefore, whilst all NWS have dual disarmament responsibilities, these are, albeit to varying degrees, much less significant than those of the US.

The next issue to consider is what method should be adopted in order to analyse the NWS, bearing all the above requirements in mind. Clearly, we need to consider state behaviour in terms of domestic and international policy, so we can have a rounded appreciation of nuclear weapons decision-making for each NWS. But we also need to recognise the different political forces at play within each NWS and neither treat the state as a monolithic entity nor mystify the process by which the ideas and materials necessary for the perpetuation of nuclear weapons systems are actively being reproduced. In addition, given the centralisation of such decision-making power in the hands of very small military and political groups across NWS, it is important to be clear that, when referring to one or more of these 'states', we mean political executives whose actions principally reflect the interests and will of private concentrations of economic, social and political power, rather than the citizenry at large.

Recognising the provisional nature of the state and state practices allows us to approach NWS behaviour, past and present, with a fresh and critical eye as required by this study given its intention to identify the potential for future, transformative, nuclear disarmament action. The approach I take to analysing NWS's nuclear weapons decision-making processes, policy choices and debates as they evolved over several decades, may be described as institutional-historical. In terms of methodology, the investigation presented in Part II is based on a critical analysis of academic, civil society, media, military and government documents, primarily from sources based in the five NWS, which variously include: historical works; policy and briefing papers; official statements; media reports; public opinion polls and surveys; speeches and debates. This analysis is enhanced by insights gathered from the 30 interviews I conducted with notable academics, campaigners, researchers and former or serving government officials from each of the five NWS.

As a point of departure, and to explain what this analytical approach entails, we may identify a few of the similarities and differences between NWS that raise interesting questions. For example, some of the NWS's governmental structures or 'regime types' have more obvious similarities than others. France, the UK and the US have formally liberal, democratic institutions with more of an elitist than popular character, whilst China and Russia are significantly less liberal, less democratic and more authoritarian. Yet all five NWS are unique in terms of history and culture. How have these different regime types, cultures and histories impacted on nuclear weapons decision-making and arguments concerning nuclear disarmament? Another way of approaching this question is to consider, as McLean and Miall have done, the similarities between processes of nuclear weapons decision-making across NWS.[6] For example, is it reasonable to conceptualise nuclear weapon 'systems' as a singularly elite, military-technical culture shared across NWS? Again, if this is the case, what does this mean for the project of, and prospects for, nuclear disarmament today?

As well as considering the factors reinforcing the continuation of nuclear weapons possession and modernisation, there is also the important question of what countervailing or opposing forces to the status quo exist, both nationally and internationally, and how and why the relative influence of these forces has changed over time? Given the meaning and requirements of *institutional democratisation,* other questions considered in more detail in the following five NWS case studies include: what do recent public opinion polls regarding nuclear weapons issues conducted in NWS, and globally, mean for the prospects of nuclear disarmament and, more broadly, what is the current status of national and global nuclear disarmament movements?

## Notes

1  Kydd, Andrew H. (2005) *Trust and Mistrust in International Relations* (Princeton, NJ: Princeton University Press), 80.
2  Ibid., 3.
3  Ibid., 4.

4 Ibid., 82.
5 Ibid., 4, 83.
6 McLean, Scilla, ed. (1986) *How Nuclear Weapons Decisions Are Made* (Basingstoke: Macmillan); Miall, Hugh (1987) *Nuclear Weapons: Who's in Charge?* (Basingstoke: Macmillan).

# 3 United States of America

## 3.1 Introduction

This chapter provides an in-depth investigation of US nuclear politics in order to assess the explanatory power of *institutional democratisation,* alongside mainstream and realist approaches, regarding the causes and consequences of US nuclear possession and disarmament. The length of this chapter is justified by the scale and global importance of the US military and nuclear establishments—which create the high level of responsibility the US has regarding the realisation of nuclear disarmament in and between NWS—and the need for a balanced analysis of US domestic and international nuclear politics.

The chapter begins with a summary of how mainstream and realist approaches relate to the US's particular experience as an NWS and how the US's nuclear status has been justified over time. For example, the US's development of nuclear weapons is placed within the context of relevant historical events from the mid to late 20th and early 21st centuries—focusing, in particular, on the Cold War. This review is based on the works of prominent military analysts, academics, historians and official documents. In such accounts, the justification for the US's development and then continued possession of a nuclear arsenal focuses on the threat posed by external enemies—beginning with Nazi Germany during WW2, the Soviet Union during the Cold War and more nebulous adversaries and opponents, including from developing nations, in the post-Cold War world.

Following this, I begin the process of showing how the US's experience as an NWS illustrates the limitations of mainstream claims and approaches, and the ways in which *institutional democratisation* helps us better understand the causes and consequences of US nuclear possession and disarmament. In doing so, I examine the domestic political drivers of US nuclear possession and the impact possessing the bomb had on the US polity. I start with the contention that the Manhattan Project marked the beginning of the US national security state and then consider the widening role the bomb played in the US's global strategy following WW2. The discussion here draws on the work of authors such as Robert Dahl, who focus on how nuclear weapons exist outside of the democratic process, to highlight arguments compatible with the main contentions of my study. I also discuss how elite actors and groups shape US defence and foreign policy, including on nuclear issues.

Having provided this comparative historical overview, I go into more detail concerning what *institutional democratisation* would mean in terms of political change for the US when reviewing modern-day US nuclear politics, thereby updating the work of Dahl and others. This includes a consideration of the short to medium term nuclear arms control, non-proliferation and disarmament measures that have been proposed, and the reasons for Washington's continued inaction and obstruction. I argue that such measures are important, but also limited in terms of how nuclear abolition might be achieved, because they often conform to a more mainstream, liberal, *guardianship* analysis of the topic that focuses on the international level and interactions between NWS elite decision-makers as the basis for cooperation. Moreover, a range of evidence is presented to show both how wide the democratic deficit is in the US, and how this is reflected in the gap between US public opinion on arms control, non-proliferation and disarmament and the actions of the US government, suggesting deeper democratic reforms are needed if more substantial disarmament action is to be taken.

I then provide an overview of the US's post-Cold War global strategy and the role played in it by nuclear weapons, discussing how the US's advanced conventional military power has a singular impact on other NWS's nuclear thinking and the implications of this for nuclear disarmament. In addition, I give an overview of the debate concerning advanced and emerging military technology, including how the US's pursuit of military dominance is affecting the threat perceptions, strategic calculations and nuclear policies of China and Russia. From here, I argue that the US has a unique dual responsibility to advance nuclear disarmament by taking progressive steps on a national level and by acting in ways that support such steps being taken by the other NWS.

Another, more recent development in the US's nuclear arsenal concerns how it is managed and reproduced by private corporations—which I discuss in terms of what this means for democracy, transparency and accountability. Subsequently, the present state of the US peace and disarmament movement is considered to explore its potential contribution to disarmament initiatives and how it may develop and be strengthened as part of a wider democratisation process. The chapter ends with a summary explaining the significance of the US case for the main arguments put forward by this study.

## 3.2 Mainstream and realist perspectives on the causes and consequences of US nuclear possession and disarmament

Despite a variety of disagreements within mainstream and realist thought, such works generally see US nuclear possession since 1945 as justifiable and necessary. Several reasons are provided to justify this stance, including: the anarchic nature of the international state system; the peace, security and stability provided by deterrence in bipolar and multipolar worlds; the liberal, benign nature of the US's global power; the need to protect US allies, values and interests; and the difficulties and dangers posed by disarmament.

For example, historian Richard Rhodes argues that the original motive of the physicists who developed the bomb for the US was fear of Germany triumphing in WW2 and establishing 'a thousand-year Reich made invulnerable with atomic bombs'.[1] The atomic bombings of Hiroshima and Nagasaki were then widely celebrated in the US for leading to Japan's early surrender. Until the Soviet acquisition of the bomb in 1949 the US held an atomic monopoly, yet initiatives to manage atomic weaponry via an international agency—including the Acheson-Lilienthal report and the Baruch plan, proposed by the US, and the Soviet Gromyko plan—failed due to mutual mistrust between the superpowers.[2]

Such mistrust meant that the Cold War quickly became a nuclear standoff between East and West. Yet there was no consensus in the US concerning what the role of nuclear weapons should be. Tom Nichols and Dana Struckman therefore observe how

> there were strong divisions among American strategists about the purpose of nuclear weapons. For some, they existed only to deter nuclear attacks on the United States; for others, they were the military equalizer between an outgunned West and a gigantic Communist alliance.[3]

The latter position informs the analysis presented by US Brigadier General Robert Spalding, who argues that the US defeated the Soviet Union to win the Cold War 'by maintaining a credible nuclear force'. According to this position, the US deployed nuclear weapons in Europe to defend it against the Soviet's 'numerically superior conventional force'. Moreover, the US's formidable nuclear triad 'deterred the Soviets from attacking' and were at the 'forefront' of the US's 'defense strategy' receiving 'priority in both rhetoric and funding'.[4]

Spalding's justification for US nuclear possession here rests on the idea that nuclear weapons are the 'US's instruments of peace' and that 'peace can only be secured through strength'. Moreover, nuclear weapons are the 'only aspect' of the US's 'national defense' that are capable of delivering such 'peace'—and do this affordably.[5] Spalding's narrative implicitly draws on a traditionalist perspective which, Joyce Kaufman explains, sees the Cold War as 'a war of *ideology,* which assumed that the two divergent approaches (democracy and communism) could not coexist peacefully. Therefore one side would have to emerge as dominant.'[6] Nuclear possession, according to this ideological interpretation of the Cold War, was thus essential for the US, if it was to defend its democracy from the threat posed by communism. Indeed, Mearsheimer has described the US's nuclear arsenal as 'an ideal middle-class weapon and a strong force for democracy'.[7] This is partly because, he claims, without a nuclear arsenal during the Cold War, the US may have had to become a 'garrison state', leading to 'crazy domestic politics' as defense spending sky-rocketed with a bulging military.[8]

Framing nuclear deterrence as both essential for the defence of US dem-ocracy and legitimate because 'approved, through the political processes of the democratic nations it protects, since at least 1950', as President Reagan's Defence Secretary Caspar Weinberger argued, is—on the face of it—a particu-larly authoritative means of justifying nuclear possession.[9] It is also one with several implications for this study, given our central contention that nuclear disarmament—albeit to different degrees across the NWS—requires *institu-tional democratisation*. Proponents of US nuclear possession today may thus point to polling data showing public support for the US's status as a NWS, and cite Kerry Herron's argument that 'the importance the general public attaches to US nuclear weapons capabilities is growing'.[10] Benjamin Valentino and Scott Sagan also argue that current attitudes amongst Americans to 'the use of nuclear weapons and the killing of noncombatants' shows that US public opinion 'is unlikely to serve as a serious constraint on any president who might consider using nuclear weapons in the crucible of war'.[11]

Despite such findings, however, realist thought and *guardianship* arguments generally tend not to ascribe great importance to public opinion, for reasons discussed in Chapter 2. Where prominent establishment figures do discuss the public, it is as a body requiring expert guidance. For example, former Defense Secretary William Perry emphasises the role of the public, stating that the US won't lead the world in 'reducing nuclear dangers … unless Americans under-stand the importance of doing so'.[12] However, Perry's argument exemplifies the *guardianship* approach to nuclear politics because he believes elites must educate and guide citizens towards prudent policy.

The US government has recently presented several arguments justifying its continued possession of nuclear weapons. For example, former Defense Secretary Robert Gates provided a succinct justification in his foreword to the 2010 Nuclear Posture Review (NPR), stating that:

> as long as nuclear weapons exist, the United States must maintain a safe, secure and effective nuclear arsenal to maintain strategic stability with other major nuclear powers, deter potential adversaries and reassure our allies and our partners of our security commitments to them.[13]

Elsewhere, nuclear deterrence has been presented by the US government as 'the ultimate protection against a nuclear attack on the United States', preventing adversaries from pursuing undesirable courses of action, such as aggression with WMD, or nuclear weapons.[14]

The US also claims that its extended deterrence relationships with regional allies (such as Japan and South Korea) provide reassurance and protect these—and other nations—from conventional or nuclear attack, as well as reducing their need to develop their own nuclear arsenals.[15] In addition to these central and extended deterrence roles, the 2014 *Quadrennial Defense*

*Review* states that the US's nuclear arsenal 'also supports our ability to project power by communicating to potential nuclear-armed adversaries that they cannot escalate their way out of failed conventional aggression'.[16]

With regards to disarmament, as we have seen, mainstream and realist approaches from US academia are particularly sceptical regarding its possibility, need and benefits. Representatives of the US government, meanwhile, have previously claimed that the US has 'an outstanding and unequalled record of compliance' with its NPT Article VI obligations, and that it continues to reduce the role and size of its nuclear arsenal.[17] In more hawkish variants of the US mainstream and establishment discourse, nuclear disarmament is explicitly rejected. For example, former senior US officials Harold Brown and John Deutch describe even the 'aspirational goal' of 'eliminating all nuclear weapons' as 'counterproductive' and a 'fantasy' since it 'risks compromising ... U.S. security and international stability'.[18]

Opponents of US nuclear disarmament also often present Russian behaviour as a key barrier to meaningful nuclear negotiations. The official US narrative of Soviet and, thereafter, Russian intransigence and obstructionism, has its origin in accounts of why early Cold War initiatives to exert international control over the bomb failed.[19] Today, US government officials frequently argue that Russia's nuclear and conventional military modernisation, as well as its recent annexation of the Crimea, threaten to derail arms control processes and strategic stability more widely. In particular, Moscow's development of new missile technology led Washington to claim that Russia is undermining 'numerous agreements', leading the US to suspend its compliance with the Intermediate-Range Nuclear Forces (INF) Treaty in February 2019.[20] Subsequently, in May 2020 the *Washington Post* reported that officials in the Trump administration had 'discussed whether to conduct the first U.S. nuclear test explosion since 1992' as a means of increasing pressure on China and Russia to sign a 'trilateral deal' regulating their nuclear arsenals.[21]

The Chinese, Iranian and North Korean nuclear programmes also often feature in recent official US pronouncements to justify why Washington is unable to make progress on disarmament. For example, at his confirmation hearing for the position of Vice Chairman of the Joint Chiefs of Staff in 2015, General Paul Selva stated that he 'would put the threats to this nation in the following order: Russia, China, Iran, North Korea'.[22] Steven Lambakis argues that each of these four nations has 'space, ballistic missile, and counter-space programs as well as nuclear and weapons of mass destruction programs that could produce systems that would allow each to engage in activities that are detrimental to U.S. security'.[23] Moreover, in May 2020, Marshall Billingslea, President Trump's arms control envoy, stated that China is 'intent on building up its nuclear forces and using those forces to try to intimidate the United States and our friends and allies'.[24]

Since becoming President, Donald Trump has alternated between threats, sanctions and diplomacy with Kim Jong-Un, North Korea's Supreme Leader, in an attempt to agree the latter's denuclearisation. However, several commentators have criticised Trump's approach, including the lack of meaningful results from the summits held between the two nations.[25] North Korea's series of intermediate-range ballistic missile tests since 2016 have been mostly unsuccessful, but are of concern to Washington because they threaten Japan, South Korea and the US Pacific territory of Guam. Pyongyang also tested two ICBMs in 2017, which analysts from the Center for Strategic and International Studies claim are 'likely' to be capable of striking the US mainland with a nuclear weapon.[26] The US therefore argues that it needs suitable capabilities to prevent North Korea—or, in future, Iran—posing a nuclear threat. In recent years, the US has developed Conventional Prompt Global Strike (CPGS) as the 'sword' for a 'counternuclear' strike against an opponent's missile bases, and Ballistic Missile Defense (BMD) as the 'shield' to intercept missile attacks.[27]

Regarding Iran, Trump has taken a particularly aggressive stance against the leadership in Tehran. In 2018 the US, with backing from Israel and Saudi Arabia, withdrew from the Joint Comprehensive Plan of Action, which was a hallmark achievement of the Obama administration. Under the agreement, Iran limited its sensitive nuclear activities and consented to international inspections in return for heavy economic sanctions being lifted. Subsequently, Tehran and Washington engaged in a heated war of words which turned increasingly violent, reaching a peak in 2020 with Trump's decision to order the assassination of Iranian Major General Qasem Soleimani. The possibility of conflict between Iran and the US continues to cast a nuclear shadow over the Middle East with Washington, alongside Tel Aviv and Riyadh, claiming that they must maintain strong military forces to prevent Iran establishing itself as the 'dominant regional power'.[28]

Moreover, the US's decision in 2020 to deploy so-called 'low-yield' nuclear warheads on one of its submarines appears intended to deter Russia, but also Iran or North Korea. This is plausible, Julian Borger explains, given the language in the 2018 NPR, which outlines how the US is prepared to use nuclear weapons in response to 'significant non-nuclear strategic attacks', including but not limited to 'attacks on US, allied or partner civilian population or infrastructure'.[29] A primary concern here, as William Arkin explains, is that such 'prompt' and 'assured' capabilities may make Trump 'more prone to accept the use of nuclear weapons'.[30]

The perceived threats to US power in East Asia and the Middle East have also led supporters of the US's nuclear status to argue that the US must retain these weapons in order to provide extended deterrence to its regional allies. Remarks by President Trump that NATO is 'obsolete', that joint US-South Korean military exercises are too 'expensive', and that US allies must invest more in their own national defence, are seen as troubling to those of a more traditional mindset, because they believe that this may lead to countries such as Japan, South Korea, Saudi Arabia, Germany—and others—developing their own nuclear deterrents.[31] If US allies in one region seek the bomb this may also cause allies in other regions to pursue their own deterrents. It is thus

An inintended consequence!

argued that maintaining a strong and credible US nuclear 'umbrella' prevents proliferation, both nuclear and non-nuclear, advances the cause of disarmament and benefits US interests. For example, Brad Roberts has explained that 'a proliferation of strike capabilities among U.S. allies and partners would heighten the perceived U.S. risk of unwanted entanglement in crisis and escalation under the nuclear shadow'.[32]

## 3.3 Critical perspectives on the causes and consequences of US nuclear possession and disarmament

In order to usefully apply the *institutional democratisation* approach to the US, it is important to provide an alternative explanation of the origins of nuclear possession in the context of 20th-century history, focusing on the meaning of the Cold War, to support our subsequent argumentation. Of the alternative explanations of the Cold War, a revisionist perspective that highlights the importance of domestic politics and considers the relative ability of the US and Soviet Union to realise their international goals given their strategic power, shall be developed below. The main point in presenting this summary is to illustrate the divergences with traditional, post-revisionist and mainstream analyses rather than provide a full historical reconstruction, both for reasons of space and to avoid repetition, given that there are several areas of factual overlap between the different perspectives.

Revisionist perspectives on the meaning of the Cold War take different forms. One striking explanation focuses on how the US and Soviet Union were both expansionist and used the Cold War as a means of inculcating fear and obedience to manage their domestic populations.[33] In order for successive US governments to be able to justify to the public the maintenance of the huge, centrally controlled, military infrastructure constructed during WW2, it was thus necessary to repeatedly invoke the threat of an aggressive Soviet Union as part of, what Chomsky argues was, a 'national security ideology for population control'.[34]

Looking more widely, as David Jablonsky explains, an 'expansive concept of US national security' was developed during and after WW2, which was linked 'to so many interdependent factors, whether political and economic or psychological and military' that 'the subjective boundaries of security pushed out further into the world, encompassing more geography and thereby more issues and problems'. Given such a broad conceptualization of security, 'developments anywhere' could be seen by Washington as having 'an automatic and direct impact on US core interests'. The result was the growing influence of military and security concerns and an all-encompassing military establishment. For Jabolonsky, this was reinforced by the Soviet military build-up, including its explosion of a nuclear device in 1949, which implanted 'the image of an external threat' in the American mind.[35]

Kaufman makes the compelling point that the Cold War was thus, in addition to being a war of ideology, 'a *political* war' whereby democracy and communism were seen as inherently 'antithetical', creating inevitable conflict.[36] In such a febrile atmosphere the possession of nuclear weaponry could thus be justified as necessary to ensure national security, whilst calls for unilateral disarmament could be presented by leading US political figures, such as Democrat Senator Stuart Symington, as 'surrendering to communism'.[37]

Following the fall of the Berlin Wall, 'new aggressors' were identified by the Pentagon, including 'rogue states' who must be deterred from using WMD against the US. This was done to justify the retention of thousands of immensely powerful nuclear warheads and, according to a 1995 study by US Strategic Command, required the projection 'to all adversaries' of an 'irrational and vindictive ... national persona'.[38] Despite significant reductions to the US nuclear arsenal under President George H. W. Bush, deeper disarmament measures and a peace dividend, which many expected to follow the end of the Cold War, did not materialise.[39] This was mainly because the people and groups occupying key positions in US economic and political institutions during the Cold War retained substantial influence and control. Furthermore, scholars such as Chomsky and Leffler argue that, since the end of WW2, from which the US emerged as the 'preponderant' world power, 'the US pursued an "imperial grand strategy" which sought to ensure "the limitation of any exercise of sovereignty" by states that might interfere with its global designs'.[40] By this reckoning, the demise of the Soviet Union removed the principal brake on the US's pursuit of global hegemony, allowing it to move from 'containment' to 'enlargement', as President Clinton's National Security Advisor—Anthony Lake—put it.[41]

This brief summary of revisionist perspectives is useful to both understand the historical context in which the US nuclear arsenal emerged and developed as well as the impact of the bomb on US society and politics. For example, the consequence of President Roosevelt's decision to acquire nuclear weapons was, for Rhodes, that a 'separate, secret state with separate sovereignty' was created, beginning with the Manhattan Project in 1939, which employed 130,000 people and was of a similar size to the entire US automobile industry.[42]

The immense size and secrecy of the US's nuclear weapons system, whereby top-level bureaucratic, military and political actors accumulated and centralised huge power away from the public gaze in the name of national security—justified as vital at a time of war—led the Atomic Heritage Foundation to describe the legacy of the Manhattan Project as the creation of the 'national security state'.[43] Gary Wills explores this idea in his work, arguing that the US's acquisition of the bomb changed US history 'down to its deepest constitutional roots' because it 'redefined the presidency, as in all respects America's "Commander in Chief"', whilst also helping to foster 'an anxiety of continuing crisis, so that society was pervasively militarized'.[44]

Crucially therefore, a wartime mentality continued into peacetime, so that the US's nuclear arsenal was purposefully insulated from accountability and transparency measures, making it much more difficult to introduce democratic controls over these weapons, an essential condition for non-proliferation and disarmament efforts. Musing on the great changes his nation had experienced, President Eisenhower used his farewell address to warn that 'public policy could itself become the captive of a scientific-technological elite' and that US 'liberties' and 'democratic processes' must be guarded against the 'acquisition of unwarranted influence, whether sought or unsought, by the military-industrial complex'.[45]

For Robert Dahl, in *Controlling Nuclear Weapons*, the bomb thereby presented a 'tragic paradox'. This was because, whilst 'no decisions can be more fateful for Americans, and for the world than decisions about nuclear weapons', such decisions 'have largely escaped the control of the democratic process'. Dahl therefore argued that nuclear possession had contributed to the 'alienation' of political control. Rather than delegating authority to responsible experts in a representative democracy of the kind envisaged by the founding fathers, the US democratic process was thus both becoming more and more 'hollow' and increasingly clothing 'a de facto regime of guardianship'.[46]

David Meyer also highlights the gap between elite and public understandings of US nuclear policy, arguing that:

> despite conflict among policy makers about the political utility of nuclear weapons, US national security policy and the role of nuclear weapons within it, has been remarkably consistent. It has also been generally isolated from wider domestic political debate.[47]

Both Meyer and Daniel Ellsberg have outlined the particular strategic benefits that US policy makers ascribe to nuclear weapons. For example, Ellsberg observes that the US made 25 'threats or consideration of nuclear first use in crises' between 1948 and 2008.[48] Meyer states that such actions can be explained by the fact that, since the US's atomic bombing of Japan, the US has 'consistently' used its nuclear force 'primarily not to protect the territorial security of the United States but to support conventional forces and foreign policy goals'. Moreover, the development of 'increasingly diverse and multifaceted nuclear capabilities' has provided the ultimate guarantee of 'US military superiority in pursuit of a wide range of political and military goals'. Crucially, for Meyer, the pursuit of these goals has meant that 'Pentagon planners and elected officials involved in making policy have necessarily conceived of the use of nuclear weapons considerably more flexibly and broadly then has the general public'.[49]

Meyer's work, in particular, complements Dahl's emphasis on the importance of democratic control concerning nuclear weapons, an idea that is explored in different ways by a range of authors. According to these analyses, domestic political concerns—including competing bureaucratic and institutional interests—are significantly responsible for driving nuclear weapons programmes. Moreover, the power of committed citizens and an engaged

Congress to halt military programmes and achieve arms control agreements is highlighted.

Other authors exploring this theme include April Carter, who wrote on the domestic obstacles to nuclear arms control, including Presidential and Congressional electoral cycles and politics, as well as the 'difficulty of keeping secrets in Washington', and Janne Nolan who discussed the domestic politics of nuclear strategy, including the associated 'entrenched bureaucracy' of military planners.[50] Elsewhere, Peter Pringle and William Arkin focused on the secrecy surrounding the development of US nuclear war plans, whilst James Lindsay considered the role of Congress in nuclear weapons decision-making.[51] More recently, Tom Sauer's work reviews the bureaucratic politics driving US nuclear weapons policy.[52] Other historical works by former senior government officials, such as Strobe Talbott, highlight the 'permanent, institutionalized acrimony' behind US policy on 'strategic arms control'.[53] Whilst not all of these authors explicitly advocate or even focus on the elimination of the US's nuclear weapons, their work helps us to identify the main domestic obstacles to nuclear arms control and disarmament, including the powerful elites and related institutions responsible for reproducing the bomb.

For example, Sauer argues that in the mid to late 1990s, there was a 'growing consensus' amongst non-governmental experts in support of minimum deterrence.[54] As advocates of this policy, such as Hans Kristensen, Robert Norris and Ivan Oelrich explain, this would involve nuclear weapons having just one mission, 'to deter the use of nuclear weapons', which would, they claim, 'lessen the legitimacy of nuclear weapons and allow for significant reductions in global stockpiles'.[55] However, for Sauer, the political power of the arms control community and the peace movement at this time—which are important contributors to any process of *institutional democratisation*—was too weak to overcome the might of the 'nuclear weapons establishment'.[56] With regard to the peace and disarmament movement, Jacqueline Cabasso also notes that during the 1990s several NGOs working on these issues moved to Washington, DC, and shifted their approach to 'securing Russian "loose nukes" and keeping them out of the hands of "rogue" states and terrorists'.[57]

As for the nuclear weapons establishment, Sauer argues that this consists of a network of 'gigantic bureaucratic organisations', such as the Pentagon, which had been created during the 1940s and 1950s.[58] These institutions had four key interests in protecting the US nuclear weapons system from fundamental change—'budget, personnel, autonomy and prestige'—which the proposed shift to minimum deterrence directly threatened. Moreover, whilst many in the US armed forces saw nuclear weapons as irrelevant and felt they should be discarded in order to procure usable capabilities, any move towards disarmament was vetoed by a small group within the military—centred around the nuclear 'targeting community'.[59]

Given the relative strength of the nuclear establishment and the weakness of the proponents of minimum deterrence—or nuclear disarmament for that matter—Sauer therefore proposes that there needs to be a 'major societal debate' in the US if the bureaucratic resistance to change is to be broken. For him, this would necessitate Presidential leadership, which harnessed the weight of public opinion in order to form a 'bipartisan consensus' in Congress.[60] In concluding her review of the failures of some members of the Clinton administration in attempting to reform nuclear weapons policy, Nolan puts forward a similar view, but focuses principally on the failure of the White House to show leadership and expend political capital in order to take on established bureaucracies—principally the Pentagon.[61]

Other critical and revisionist analysis focuses on the significant popular support in the US for action pursuant to nuclear disarmament. According to this perspective, presented, for example, by Cortright, key events—such as the Reykjavik nuclear arms reduction summit involving Reagan and Gorbachev—therefore need to be seen in relation to the popular pressure in East and West to avoid a nuclear conflagration.[62] Jeffrey Knopf also uses a 'domestic structure approach' to show how US 'citizen activism on behalf of arms control and disarmament' did have a significant impact on Washington's 'preferences for cooperation' and entry into 'strategic arms talks'.[63]

The other side of this coin, for authors such as Meyer, is that US decision-making elites have historically co-opted, sidelined or ignored underlying support for nuclear arms control, non-proliferation and disarmament amongst the populace.[64] Such trends continue to this day, so that a 2004 poll by the Program on International Policy Attitudes found that a majority of the US public was 'not aware' that the US had 'made a commitment to seek the ultimate elimination of nuclear weapons as part of the Non-Proliferation Treaty'. When made aware of this fact, however, 'a very large majority' thought that 'doing so was a good idea and that the US should make greater efforts toward that goal'.[65] This finding is again representative of the broader absence of public debate concerning nuclear weapons policy in the US. Despite this, a 2009 report on how to communicate nuclear weapons issues to the public by the US in the World initiative argued that

> the fact that the public does not think about nuclear weapons issues—yet still supports deep reductions in the number of weapons, and in some cases, concludes on their own that we need to eliminate all nuclear weapons is a real opportunity for advocates.[66]

Thus, contrary to those, such as Christopher Ford, who argue that the US's adherence to Article VI of the NPT is 'exemplary', critics point to a series of actions the US should immediately take to begin reducing the salience of nuclear weapons in its national security policy and realise its dual international disarmament obligations.[67] Such measures and proposals include: ratification of the CTBT; negotiation of a FMCT; support for all five existing and other proposed nuclear weapon free zones, including the Middle East WMD Free Zone; and the removal and destruction of the so-called 'tactical' weapons the US deploys in five NATO nations, under 'nuclear sharing arrangements'.[68]

Other early proposed actions, specifically directed at changing the policies governing the US's nuclear weapons, include: taking nuclear weapons off alert; adopting a policy of No First Use; and the retirement and verified elimination of non-deployed reserve stockpile weapons.[69] There are also a number of action supportive of disarmament that the US could take in relation to diplomacy, arms control and trade, including Washington's ongoing support for the nuclear weapons programmes of NPS, such as the UK.

Moreover, critics of US nuclear policy argue that recent changes wrought by arms control agreements such as New START (Strategic Arms Reduction Treaty) do not match the majority of the public's expressed desires or expectations. Indeed, commenting on their research into US and Russian public opinion on nuclear arms control and disarmament, John Steinbrunner and Nancy Gallagher observe that 'responses to detailed questions reveal a striking disparity between what U.S. and Russian leaders are doing and what their publics desire'. Steinbrunner and Gallagher therefore suggest that US political leaders should be much bolder in using their 'bully pulpits to solidify and mobilize public support' in order to articulate 'a compelling alternative that is more in line with the public's core values'. One of the broad alternatives these authors outline based on their research is that the US (and Russian) people would support leaders 'who directed their own bureaucracies to alter fundamentally both the guiding objective and the action program used to address the challenges of the new nuclear era'.[70]

Another important aspect of the domestic US debate regarding the need for a strong military and nuclear force concerns the threat posed by North Korea and Iran to the US, particularly given these two states' nuclear ambitions. It is clear that weaker powers may choose to pursue a nuclear weapon programme to counter more powerful state's conventional capabilities as part of their strategies for central deterrence and regime survival. Alan J. Kuperman thus observes that NATO's bombing and overthrow of Libyan leader Muammar Gaddafi 'greatly complicated the task of persuading other states to halt or reverse their nuclear programs'.[71] The lesson Tehran and Pyongyang learned from this episode is that, because Gaddafi had voluntarily ended his nuclear and chemical weapons programmes, the West now felt free to pursue regime change.

However, those in the US arguing for military action against Iran and North Korea often frame these nations, to borrow terms from Becker, Müller and Wisotzki, as being 'unjust' and 'undemocratic' enemies.[72] Such framing allows the US to make a distinction between 'good' and 'bad' nuclear possessors, which must, respectively, be protected, controlled—or even eliminated. This is significant in terms of the claims made by *institutional democratisation* because, as the above authors note, in democracies such as Canada and Germany, 'the concept of the unjust enemy is virtually nonexistent', whilst in Ireland 'the images of the undemocratic other and rogue state are completely absent'.[73]

As these authors argue, we therefore need to consider the 'special role' that the US has adopted, as 'superpower and protector of world order', in order to understand how and why it makes the nuclear and other strategic choices it does.[74] For the purposes of this study, such an analysis should be extended to consider how the US's role and behaviour might change in ways which are compatible with nuclear disarmament. The following sections consider this question with a particular focus on the domestic political forces driving nuclear weapons decision-making in the US, and the practical importance of *institutional democratisation* for efforts towards eliminating the US nuclear arsenal.

## 3.4 US nuclear politics in the post-Cold War world

The previous section explored both the gaps in mainstream and realist accounts of the causes and consequences of US nuclear possession and disarmament and summarised the explanatory value of *institutional democratisation* for this study in relation to the historical record. This section builds on these insights concerning the value of *institutional democratisation*, first, by examining the role nuclear weapons occupy within the US's global strategy in the post-Cold War era. Understanding the US's strategic goals—and what domestic actors and groups drive and shape these goals, both historically and in recent times—is vital if we are to properly examine the political barriers preventing the US moving to low numbers of nuclear weapons on the path to zero. Furthermore, comprehending the benefits decision-making elites believe nuclear possession brings—domestically and internationally—is an important part of understanding their resistance to disarmament.

The discussion then moves on to explore in more depth the domestic politics that shape Washington's strategic behaviour. For example, we shall examine how the nuclear weapons establishment is embedded within the US economy, state and society, and the obstacles to and opportunities for progressive change supportive of disarmament, such as *institutional democratisation*, presented by the status quo. In terms of obstacles, this means outlining which powerful domestic actors and groups, with economic and political influence over nuclear weapons decision-making, drive the US's continued nuclear possession and prevent disarmament action. As for opportunities, this includes identifying the current state of those actors and groups in government and civil society that endorse progressive action supportive of nuclear disarmament. In addition, given the US's immense conventional military power—which has a singular impact on the strategic thought of all other states—the ideas of *institutional democratisation* can and should also be applied to US defence and foreign policy in general to imagine a process by which Washington reorients its global strategy so that it meaningfully contributes to and supports other NWS realising their nuclear disarmament obligations.

Before beginning this analysis, it is useful to briefly consider the key governmental actors involved in US nuclear weapons decision-making, including their role and position in the policy hierarchy. Ritchie provides a detailed overview of this in his study of US nuclear weapons policy after the Cold War, stating that 'the White House, in particular the office of the president, and the National Security Council' sits at the centre of decision-making, with the 'next policy ring encompassing executive departments and agencies'. Outside the executive and in the next 'policy ring' lies Congress, consisting of the Senate, House of Representatives and the judiciary.[75] As Lindsay notes in his study *Congress and Nuclear Weapons*, 'members of Congress, no matter how well-intentioned, almost always lack in-depth understanding of nuclear issues', so that Congress fits within Dahl's *guardianship* model, given the 'tremendous disparity' between it and the executive's nuclear knowledge.[76]

Given the primacy of the Presidency in these matters, for example, concerning nuclear use decisions and the power to make unilateral reductions to the nuclear arsenal without seeking Congressional approval, when reviewing the post-Cold War history of US nuclear weapons policy, it appears reasonable to organise our analysis of the post-1989 period according to Presidential administrations. The problem with following this approach is that whilst, formally speaking, the principal responsibility for nuclear weapons policy lies with the President, as McLean's study *How Nuclear Weapons Decisions are Made* argues, in practice it is the power of the 'permanent bureaucracies' which both sets the agenda and strongly mitigates against change. Indeed, bureaucrats across NWS control the development of nuclear weapons systems which now last for 'fifteen or twenty years', much longer than the period in which government ministers—even US Presidents—are in office.[77] There is thus great pressure on elected officials to maintain consistency with their predecessors' decisions, reducing opportunities and space for democratic deliberation and participation.

As McLean's study explains, whilst the US President is the 'ultimate decision-maker' on nuclear weapons issues, with the responsibility for implementing them, he or she will be faced by significant limitations, because 'the President will rarely see options unless he insists on them; by the time he is presented with a weapons issue for approval, almost all of the decisions are made'.[78] Faced with an immensely powerful bureaucracy with deep-rooted interests in the status quo, unless the President has a 'comprehensive, alternative policy formulation for national security', making a 'single system disapproval or variation' would therefore have 'no logic'.[79] The following sections, which review post-Cold War US nuclear politics sequentially across presidential administrations, should thus be read with these caveats in mind.

### 3.4.1 The Presidencies of George H. W. Bush, Bill Clinton and George W. Bush

In *Empire and the Bomb*, Joseph Gerson explains how 'as the end of the Cold War began to be anticipated', the Reagan administration brought together senior military and strategic planners to formulate proposals for the new era.[80] This resulted in the *Discriminate Deterrence* report of 1988, which highlighted that 'Japan and Europe were beginning to challenge US global hegemony' so that the US needed to focus on three regions: the Persian Gulf, Mediterranean and the Pacific Ocean, in order to remain the world's dominant power. Moreover, this strategy necessitated that the Pentagon prioritise the modernisation of its nuclear arsenal and invest in military capabilities for rapid military intervention, including high-tech weaponry. Gerson points out that the 1991 Gulf War saw this report's 'rationales and strategy' being put into practice, a war which also saw London and Washington issuing nuclear threats to Iraq.[81]

Behind closed doors, military planners continued to develop ambitious plans for how nuclear weapons could be used. According to William Burr, the leaked 1992 *Defense Planning Guidance* showed how, during the George H. W. Bush administration, Pentagon officials 'tried to develop a strategy for maintaining U.S. preponderance in the new post-Cold War, post-Soviet era', key to which was 'preventing the reemergence of a new rival'.[82] One of the authors of the report, Andrew Hoehn, argued that the US 'must continue to maintain a diverse mix of survivable and highly capable nuclear forces, including non-strategic nuclear forces', which, as Burr explains, would support the US's 'global role', 'validate security guarantees' to regional allies and 'deter Russian nuclear forces'.[83]

Thus, as Arkin described, even before the end of the Cold War—as the nuclear complex grew 'idle' and a test ban loomed—'nuclear advocates in the military and the laboratories began highly creative efforts to identify new "requirements" for nuclear weapons'.[84] These efforts were partly a response to fears that new facilities would be cancelled and existing sites closed—causing the loss of key production capabilities and leading to 'structural disarmament'.[85] Proposals thus emerged for a new generation of weapons including 'mini-nukes'—so-called 'low-yield' warheads capable of penetrating the ground for 'hard target and surface attacks'.[86] Arkin warned that 'just conducting research' on these weapons would have several damaging consequences for international cooperation on security. For example, US–Russia relations would be harmed, 'anti-democratic military and nuclear mafias in Russia' would be strengthened whilst non-proliferation efforts and the chances of agreeing a test ban would be seriously undermined.[87]

The reshaping of the US's nuclear arsenal after the Cold War also involved a 40% cut to the nuclear weapons stockpile, whilst the Presidential Nuclear Initiatives included the removal of the US's forward deployed tactical nuclear weapons and reductions to the launch readiness of the nuclear arsenal. These moves led the Soviet Union/Russia to make reciprocal measures and cuts to its nuclear arsenal.[88] Yet, as Ritchie notes, the US's moves were a 'pragmatic response to geo-political, financial and technical realities rather than a sweeping away of Cold War doctrine'.[89]

Moving on to President Clinton's term in office, Kristensen observes that, whilst Clinton pledged that the US would not use nuclear weapons against

NNWS, he also expanded its nuclear war plans to 'take on a broader role including rogue states armed with weapons of mass destruction'.[90] Several observers of Clinton's policies from quite different backgrounds, including General George Lee Butler, nuclear weapons policy expert Janne Nolan and disarmament activist Jacqueline Cabasso, therefore criticised his administration for squandering the opportunity after the Cold War to enact reforms that would delegitimise nuclear weapons.[91] Instead, as Arkin and Sauer argue, Clinton made a series of decisions that strengthened and 're-legitimised' the US nuclear weapons system.[92] Whilst Republicans were primarily responsible for the US not ratifying the CTBT in 1999, Sauer notes that Clinton's campaign to ensure it passed in the Senate was a failure, showing a lack of leadership on this vital issue.[93] Overall, these developments have led Kristensen to observe that military planners—rather than democratically mandated political leaders—had achieved significant 'leverage' in shaping the US's nuclear policies after the Cold War.[94]

The polarising nature of President George W. Bush's administration led to it being criticised for comprising a radically aggressive shift in US defence and foreign policy. Yet, as Leffler notes, Bush's policies were 'not as radical a departure from his predecessors as both critics and supporters proclaim'.[95] Rather, as Burr points out, the 1992 *Defense Planning Guidance,* calling on the US to 'prevent the reemergence of a new rival', informed the Clinton administration's strategy and foreshadowed the 'preemptive doctrine that George W. Bush has tried to turn into an axiom of U.S. policy'.[96] Indeed, as James Mann identified, the invasion of Iraq in 2003 'was carried out in pursuit of a larger vision of using America's overwhelming military power to shape the future'.[97] Furthermore, military planners during the Clinton administration signalled their determination to establish programmes ensuring 'full spectrum dominance', as outlined in the Joint Chiefs of Staff's 2000 'conceptual template' for the US's armed forces, titled *Joint Vision 2020.*[98]

With expansionist plans such as these high on the US military establishment's agenda, the stage was therefore well set for the incoming Bush administration in 2001. G. John Ikenberry summarised Washington's approach during this period as a 'neoimperial grand strategy', which, he lamented, would 'rend the fabric of the international community and political partnerships precisely at a time when that community and those partnerships are urgently needed'.[99] Elsewhere, James Goodby argued that the US's 2002 National Security Strategy is 'essential' to an understanding of the Bush administration's approach to nuclear weapons policy and also 'codified Bush's preventive war thinking', which entailed 'unchallenged military supremacy over any other nation in the world, and anticipatory military action against perceived gathering threats'.[100] In practice, this new strategic framework included the construction and deployment of BMD, as a result of which, the US unilaterally withdrew from the Anti-Ballistic Missile Treaty (ABMT) in 2002. This treaty was seen internationally as a 'cornerstone of

strategic stability' because it facilitated later agreements limiting and reducing US and Russian deployed strategic nuclear arsenals.[101]

### 3.4.2 From Barack Obama to Donald Trump

Assuming office in 2009, President Barack Obama's inheritance from his
predecessor included, according to Lawrence Korb, Laura Conley and Alex
Rothman, a 'defense budget far in excess of even Ronald Reagan's peak Cold
War spending'.[102] Regarding Obama's approach to nuclear issues, Marco
Lyons, a US Army strategist, argues that the 2010 Nuclear Posture Review
was produced as part of 'significant continuity in policy and posture since
the last NPR' so that it 'reaffirmed a fundamental role for nuclear weapons
in national security'.[103] Despite this, Obama's high-profile Prague speech,
which took place in the first few months of his Presidency, where he stated
'America's commitment to seek the peace and security of a world without
nuclear weapons', was widely lauded as signalling a break from the policies
of his predecessor. Indeed, Obama's promotion of a NWFW was cited by
the Nobel Committee as a reason for awarding him that year's Nobel Peace
Prize.[104] However, the nuclear policies Obama pursued were largely in harmony with preceding administrations, being shared, as Joe Cirincione avers,
by US national security experts 'across the political spectrum'.[105]

Further evidence of this continuity was provided in 2012, when the
*Washington Post* reported that the US's nuclear arsenal was 'set to undergo
the costliest overhaul in its history'.[106] The DOD and Department of Energy
(DOE) plan to spend approximately $1 trillion over the next 30 years, including
$68 billion to develop and purchase a new generation of nuclear bombers,
$347 billion to purchase and operate 12 new ballistic missile submarines
and billions more on new submarine launched ballistic missiles and inter-
continental ballistic missiles, new 'low yield' nuclear weapons and supporting
facilities.[107]

These huge proposed outlays help us accurately assess President Obama's
record on nuclear weapons. For example, Lichterman argues that the spending
committed to the nuclear weapons complex by the Obama administration was
part of a political 'bargain' made with Republicans. This ensured that the New
START treaty would pass in the Senate whilst also not disturbing the development of BMD and other advanced conventional weapons programmes.[108]
Elsewhere, Michael Izbicki of the US Naval Submarine School provided
an apposite summary of how Obama's policies would damage nuclear non-
proliferation when stating that 'it will be difficult for foreign powers to conclude that the United States is serious about a long-term reduction in nuclear
weapons while we are modernizing our infrastructure so dramatically'.[109]

According to Ben Zala and Andrew Futter, Obama's administration also
sought to 'increase the role of advanced conventional weaponry' in order to
'reduce its own nuclear stockpile'. Yet these authors highlight the danger that
magnifying US conventional superiority 'essentially works to decrease US

vulnerability in a nuclear disarmed world, while at the same time increasing the vulnerability of its current or future rivals and adversaries'.[110] For example, China and Russia see their nuclear weapons as a means of deterring the threat posed by the US's far superior conventional forces, including in their strategic calculations the US's advanced military capabilities, such as BMD, precision-guided munitions, long-range conventionally armed weapons that can be assigned strategic goals and its weaponisation of outer space.[111] Perkovich and Acton therefore make the important point that 'an eventual nuclear-abolition project' would require the US to give assurances that the global elimination of nuclear arsenals will not lead to an increase in its relative military power and that it would 'abide by international law as understood by other major powers in determining whether, when and how to use military force'.[112]

President Obama's rhetorical focus on nuclear disarmament and a NWFW has, under the Trump administration, given way to a focus on re-establishing US military superiority, thus dealing even discussion of strategic restraint, curbs on DOD spending and nuclear abolition a further severe blow. For example, under President Trump the US is advancing a militarisa-tion programme covering both nuclear and conventional weapons, as well as increasing spending on BMD by 25%, so that it was close to $10 billion for 2019.[113] In addition, the 2018 NPR extended the circumstances in which the US might respond to a non-nuclear attack—such as a cyber-attack on critical infrastructure—with a nuclear strike.[114]

Looking beyond the nuclear realm, in order to maintain its leading military-technological position, the US government is pursuing a 'third offset strategy', investing $18 billion in innovative technologies to 'sus-tain and advance America's military dominance for the 21st century'.[115] As Warren Chin explains, the third offset focuses on 'the intention to exploit advances in machine autonomy, AI, quantum computing and enhanced digital communications to improve the man–machine interface in the future battlespace'.[116] The US DOD's drive to 'build a more lethal force' needs to be understood in relation to Washington's strategic outlook. For example, in the US's 2017 NSS, Russia and China were presented as 'revisionist powers' intent on undermining the US-led international order.[117] The DOD's 2018 National Defense Strategy then identified 'long-term strategic competitions with China and Russia' as the 'principal priorities for the Department'. As a result, an 'increased and sustained investment' in military programmes was deemed essential if the US were to be able to prevail in an all-out war with one or both of these nations.[118]

Moreover, as Luis Simon observes, the last two decades have seen several countries, most significantly China, Russia and Iran, developing anti-access and area denial (A2/AD) capabilities, including 'ballistic and cruise missiles, offensive cyber-weapons, electronic warfare'.[119] These capabilities, Simon explains, 'undermine' the 'foundations of US power projection and global military-technological supremacy'.[120] The advent of A2/AD systems—and the US's restructuring and reduction of its overseas bases—has, for Amy

Woolf, led analysts and military officials to emphasise the need for the US to 'maintain and enhance its long-range strike capability'.[121] Several systems have been considered by the US DOD to provide such a capability, including 'bombers, cruise missiles, ballistic missiles, and boost-glide technologies that would mate a rocket booster with a hypersonic glide vehicle'.[122]

The main reason for the US to develop hypersonic weapons, Dean Wilkening argues, is to 'hold Russian and Chinese mobile targets at risk and to improve the ability to penetrate advanced integrated air-defence systems'.[123] However, hypersonic weapons, particularly when armed with conventional munitions, he argues, 'could have a profound effect on strategic stability'.[124] This is because the speed and accuracy of conventionally armed hypersonic weapons may threaten the survival of Chinese and Russian mobile ICBMs, which are 'the backbone' of these nations' 'land-based strategic nuclear forces and their respective nuclear deterrents'.[125] Overall, for Wilkening, the development of hypersonic weapons means that 'misperception, misunderstanding and miscommunication in the midst of war are more likely, contributing to inadvertent escalation', yet, hitherto, 'suggested approaches to avoiding their destabilising effects do not appear promising'.[126]

As Wilkening and several other scholars note, hypersonic weapons are but one part of a wider trend towards greater speed in contemporary military affairs, made possible by technological developments, including in anti-satellite weapons, cyber and artificial intelligence (AI).[127] For Futter, nuclear weapons thus now exist in a 'cyber context' which includes 'the security implications of a more extensively digitized global nuclear order and a greater reliance on information technology throughout the nuclear weapons enterprise'.[128] Cimbala has identified several reasons why we should expect 'overlap' between nuclear deterrence and cyber conflict. These include: i) future regional nuclear arms races will see competition in information technology and 'other aspects of advanced conventional command-control and precision strike systems'; ii) cyber-attacks may be used against an opponent's nuclear command and control systems, military capabilities and infrastructure during a crisis or war; iii) cyber capabilities enable escalation dominance or escalation control.[129]

Regarding nuclear command and control—referred to as NC3—the US and other nations' nuclear modernisation will thus inevitably bring their systems within closer reach of hackers. Moreover, the threat posed by cyber is particularly acute for the US nuclear arsenal since, as Erik Gartzke and John Lindsay note, NC3 has 'long been recognized as the weakest link in the US nuclear enterprise'.[130] Acton has also highlighted how the risks of inadvertent escalation to nuclear conflict are increasing because of the 'entanglement' of 'nuclear command, control, communication, and intelligence capabilities (C3I)' owned by the US, and also China and Russia.[131] The central problem Acton identifies is that an attack on these state's nuclear or C3I assets may not be 'deliberate' but the targeted state may respond using deterrent or other pre-emptive actions that are 'highly escalatory'.[132] As for AI, whilst such capabilities are yet to be fully developed, Kenneth Payne argues that functioning and effective systems would enhance 'the possibilities for successful first strike against adversaries possessing limited nuclear arsenals, and could even shift the balance against adversaries that are better endowed with nuclear weapons'.[133] Payne thus identifies the key issue with AI as who will develop the 'best algorithm', since 'marginal quality might prove totally decisive' against 'inferior rivals'.[134]

Overall, the picture concerning emerging military technology is thus a highly complicated one, raising a range of new challenges and uncertainties which have important implications concerning nuclear weapons for both the US and the other NWS. As discussed above, several scholars have raised concerns about the impact of advanced military capabilities on strategic stability and the potential for escalation to nuclear war. The trend towards militarisation, modernisation and international tension—particularly concerning the US, Russia and China—clearly bodes ill for the prospects of multilateral progress on nuclear arms control, non-proliferation and disarmament. As Lionel Fatton observes, arms control may impact on domestic politics by reducing 'the influence of the military over foreign policy'.[135] The other side of this coin is that the more successful those constituencies pushing for militaristic postures are, the weaker the influence of those advocating more restrained strategies. The next section discusses these dynamics in relation to the current state of the nuclear and conventional arms control, non-proliferation and disarmament regimes, with a focus on the US.

### 3.4.3  *The US's dual nuclear disarmament responsibility*

In order to examine in more detail the question of *institutional democratisation* in relation to the US's national nuclear disarmament obligations, we need to recognise the fact that the US has a particularly significant dual nuclear disarmament responsibility. This responsibility first relates to the US's nuclear arsenal and, secondly, to the US's immensely powerful conventional military which has a singular influence on all state's strategic thought, including as the main global driver of nuclear proliferation. Similarly, Brown and Deutch note that, 'even in the absence of overwhelming superiority in nuclear weapons, the great predominance of U.S. conventional forces would remain a strong motive for aspiring states to seek nuclear weapons'.[136] For the purposes of this study we therefore need to consider how the US can move from its current strategic posture to one supportive of NWS nuclear disarmament, and the value of *institutional democratisation* to realising this transition. I do this below by reviewing the US's current position regarding arms control and disarmament agreements, first covering nuclear weapons and then conventional weapons.

In 2015, following the Ukraine crisis, Russian nuclear expert Alexei Arbatov noted that 'nearly all negotiations on nuclear arms reduction and nonproliferation' between Washington and Moscow 'have come to a stop, while existing treaty structures are eroding due to political and military-technological developments and may collapse in the near future'.[137] The subsequent US abandonment of the INF meant the loss of a treaty that provided important verification and on-site monitoring of nuclear weapons, ensuring the elimination of prohibited items. This situation also threatens the survivability of New START, under which the US and Russia agreed to make further deep-cut commitments to their nuclear arsenals, and which expires in 2021. New START had, as Steven Pifer points out, been a 'bright spot' amidst deteriorating US–Russia relations, yet without it there will be no legally binding limits on the two nation's nuclear arsenals, nor the existing inspection and verification regime.[138]

Returning to Arbatov, in recent years he has made a series of striking interventions in the debate concerning US–Russia strategic relations, providing an essential overview of the challenges facing nuclear arms control. First, he argues that whilst several studies have proposed 'substitutes' for the INF and New START treaties, all of the presented options are 'considerably less effective than existing arms-control treaties with respect to preserving strategic stability and predictability, and managing the arms race'.[139] It would thus be much easier and more efficient to preserve both treaties, rather than looking for replacements after they have disappeared.

Secondly, writing in 2017 on US–Russia relations, he reflects that, following a six-year hiatus in arms-control talks, 'the two nations are as wide and dangerously apart as in the early 1980s regarding their understanding of the role of nuclear weapons, nuclear deterrence and strategic stability'.[140] Furthermore, this hiatus 'removed an important channel of strategic communication between Russian and American national-command authorities'.[141] Thirdly, Arbatov identifies the main problem with the strategic relationships between the major powers as neither lying in their 'technical complexity' nor the 'degree of turmoil in the world order', but rather the 'distinct failure on the part of the new generation of political and military elites on both sides to appreciate the high strategic importance and priority of arms control'.[142]

Fourthly, given the current predicament, he identifies several 'urgent measures' which are required to salvage nuclear arms control, including:

> settling controversies over alleged violations of the INF Treaty; resuming, as soon as possible, negotiations on a follow-on to New START; ensuring adherence to the terms of the Comprehensive Test-Ban Treaty; and reviving cooperation on the safety and security of nuclear materials, sites and technologies. This should be done through regular military and civilian dialogue, translated into follow-on arms-control treaties with elaborate qualitative and structural limitations on the model of START I, and enhanced by comprehensive confidence-building measures related to strategic offensive and defensive arms.[143]

In addition, Arbatov calls for the US and Russia to make a joint declaration 'excluding any nuclear first strike or first use' concerning 'winning and fighting nuclear war'.[144]

The dire state of nuclear arms control makes it all the more important for there to be robust conventional arms control measures in place, yet in this area there is the added problem that much novel military technology requires appropriate regimes to be built from scratch. Notwithstanding the understandable anxiety and pessimism that the many new technological developments may instil in the observer, it is important to recognise that the future remains open and subject to human choice, so that arms racing and conflict is not inevitable or determined by such advances. Chin makes the important related point that 'capability should not be equated with intent, and people rarely decide to initiate violence without cause. For this reason, it is important to reflect on the political context, which will provide the policy logic for war in the future.'[145] Similarly, Caitlin Talmadge observes that, rather than 'emerging technologies' existing as an 'independent, primary driver of otherwise avoidable escalation', such technology 'more likely functions as an intervening variable—a sometimes necessary, but rarely sufficient, condition for escalation'.[146] Talmadge also argues that because 'political and strategic choices' will 'shape the impact of technology' we should be realistic about what arms control can achieve in this context, and that it may not be able to restrain escalatory pressures.[147]

If we accept the limitations of arms control alone to achieving a more stable and peaceful international outlook, it is necessary that we identify alternative approaches which may be more effective. As previously noted, these could include—for the US—prioritising strategic cooperation and restraint, including rethinking its regional security relationships, and, domestically, pursuing measures supportive of *institutional democratisation,* to support efforts to demilitarise and eliminate its nuclear arsenal. Notwithstanding such far-reaching changes, analysts and experts have suggested a range of more immediate conventional arms control measures which the US could support, thereby strengthening the nuclear arms control, non-proliferation and disarmament regime.

Beginning with advanced conventional missile capabilities, Joshua Pollack argues that 'new confidence-building measures and expanded mutual transparency are warranted to avoid creating new dangers'.[148] Similarly, Arbatov has emphasised the need for 'much deeper and unbiased professional discussion' between Russia and US 'military and civilian experts', given the prospective deployment of BMD and 'highly accurate nuclear and conventional offensive systems'.[149] Elsewhere, Scheber and Guthe highlight the fact that only one of the five CPGS 'concepts' they identify would be covered by 'existing arms control constraints, namely the New START ceilings on deployed delivery vehicles and warheads'.[150] This provides another reason for the US, Russia and other relevant parties to explore how existing treaties may be revived and extended. For example, Wilkening argues that strategic stability would benefit from a ban on all 'short-time-of-flight counterforce weapons, including ballistic missiles'.[151]

In terms of the 'cyber–nuclear nexus', for Futter, identifying arms control measures to address this issue is particularly challenging because of the secrecy surrounding state capabilities.[152] He therefore proposes that governments, including those of nuclear-armed states, take the time to understand the issue, develop appropriate defences, ensuring nuclear systems are 'simple, safe and secure', and focus on confidence-building measures as the basis for wider agreements in future.[153] For Marchant and Allenby, meanwhile, the fact that cyber technology does not support the 'necessary definitional rigor that an enforceable treaty would require', and the 'fundamental divergence' between the strategies of China, Russia and the US—who are most involved in cyberconflict—is evidently problematic.[154]

Despite this, these authors argue that 'some controls' may be possible. Those they believe 'most likely to be effective, at least in the short run', for both cyber and other emerging technologies, fall under the concept of 'soft law', which they describe as 'instruments or arrangements that create substantive expectations that are not directly enforceable'. Soft law is thus distinct from 'hard law' requirements, for example, 'treaties and statutes'.[155] For Altmann and Sauer, however, 'alternatives to a ban' on autonomous weapon systems (AWS) such as a moratorium, code of conduct or major power agreement on use, are inadequate compared to a 'a legally binding, preventive, multilateral arms-control agreement comprehensively prohibiting the deployment and use of AWS'.[156]

Yet Payne argues that 'the idea of arms control for AI remains in its infancy', so that whilst civil society groups such as the International Committee of the Red Cross have 'published advisory guidance on the use of autonomous weapons', the relevant 'customary and formal international law remains in flux'.[157] Hope for progress on this issue may lie with those states meeting under the auspices of the Convention on Certain Conventional Weapons. In 2019 this group adopted a set of 'Guiding Principles' which had been 'affirmed by the Group of Governmental Experts on Emerging Technologies in the Area of Lethal Autonomous Weapons Systems'.[158]

The literature reviewed above illustrates the importance of understanding both the differences and connections between established nuclear arms control regimes and those for emerging conventional military technology, and the type of minimal steps Washington needs to take to ensure common global security. Regarding nuclear policy, as Arbatov notes, it is essential for US and Russian decision-makers to recommit to arms control, preserving and strengthening existing treaties rather than reinventing the wheel, and renounce the possibility of trying to fight and win a nuclear war. As for advanced conventional weapons, there needs to be renewed cooperation between civil society representatives, academics, experts, parliamentarians and government and military officials—across NWS—to construct appropriate new regimes and rules of the road. In order to function optimally, future arms control and disarmament agreements also need to be based on an understanding of how military technologies relate and are entangled. Such efforts would be supported, at the domestic level, by increased civilian and democratic control and oversight of decision-making concerning conventional and nuclear weapons. Given the importance of US behaviour in this area, the next section considers in more detail the domestic obstacles and opportunities to the US pursuing progressive arms control and disarmament measures in the near term.

## 3.5 Domestic actors and interests driving US nuclear weapons decision-making

In addition to a critical analysis of nuclear weapons policy at a governmental level, and in relation to the US's global strategy, we need to examine what other domestic non-governmental actors and groups are driving nuclear weapons policy. This section will first consider the different ways that analysts sympathetic to or supportive of disarmament have conceptualised and critiqued the US's nuclear weapons establishment, including the domestic economic, social and political drivers of nuclear possession they identify. I then move on to discuss the current health of US democracy and the state of prominent pro-disarmament actors and groups, both as a means of developing the concept of *institutional democratisation* in relation to the range of US institutions involved in the management and reproduction of nuclear weapons and to imagine how the current political obstacles to disarmament may be overcome.

Beginning with pro-disarmament critiques of the US nuclear weapons establishment, one prominent approach argues that the power and privilege of the US's military-industrial complex (MIC) needs to be confronted and diminished if the US is to make progress on disarmament. For example, the Women's International League for Peace and Freedom define the MIC as the

> policy and monetary relationships between legislators, national armed forces, and the so-called 'defence' industry (aka war profiteers). These relationships include political contributions, political approval for expenditure on weapons and war, lobbying to support bureaucracies, and oversight of the industry.[159]

Lichterman argues that the various actors and groups forming the MIC in the US have strong economic, political and ideological interests in maintaining the nation's nuclear weapons system. He notes that the 'still-considerable economic and political power of the immense nuclear weapons complex and associated elements of the aerospace-military-industrial complex' forms 'a national web of institutions that continues to deploy an array of ideological and institutional techniques to sustain their flow of tax dollars'.[160] Scholars such as Noam Chomsky have outlined the wider significance of the huge opportunity costs of nuclear weapons. For Chomsky and other critics, such as Chalmers Johnson, the production of these weapons is a prime example of 'military Keynesianism', which involves the state heavily subsidising private industry with public money for the production of, in this case, military hardware—which it then buys—as the sole consumer in the case of nuclear weapons.[161]

Indeed, Lichterman describes how military spending has been 'one of the few forms of government industrial policy capable of gaining any consistent consensus' across the US political system.[162] The US nuclear weapon system, as part of the MIC, or what Chomsky prefers to call the 'Pentagon system', is thus criticised on the grounds that it not only socialises the cost and risk of developing hi-tech military equipment whilst privatising profits, but also robs the US people of vital resources that could be invested in goods, services and infrastructure fulfilling basic human needs.[163]

Elsewhere, Greg Mello and William Hartung have produced analyses to explain nuclear weapons decision-making and the forces behind rocketing military spending. As Hartung states, for example, making 'sensible cuts' to the US's nuclear arsenal will require policy makers to 'take on the money, power and influence of the nuclear weapons lobby'.[164] His 2012 study, *Bombs versus Budgets: Inside the Nuclear Weapons Lobby* explains how the main nuclear weapons contractors in the US give large sums of money to members of Congress who sit on 'the four key subcommittees with jurisdiction over nuclear weapons spending'. The purpose of these donations, Hartung suggests, is in order for the nuclear weapons lobby to 'either collaborate to promote higher nuclear weapons spending or compete for their share of a shrinking pie'. The US's continued possession of and investment in a sizeable nuclear arsenal is of particular importance to these companies as they have become, to varying degrees, 'dependent' on such government contracts for income.[165]

Hartung's study is very valuable in explaining how corporate interests influence Congress in order to gain lucrative nuclear weapons-related contracts. In terms of the hierarchy of power on these issues, however, it is important to note that Congress, whilst an important player, has a limited reach. For example, Ritchie states that, despite having the 'power of the purse' so that it can remove or increase funding for a programme, Congress should be considered a 'junior partner in nuclear weapons policy-making'. He goes on to argue that this is because

> the administration dominates nuclear weapons policy by setting the agenda through its control over information and expertise and relatively few members of Congress have considerable knowledge of and interest in nuclear weapons issues.[166]

Others working on this topic, such as Mello and Damon Hill, describe how the DOE—which is the 'present-day landlord of the U.S. nuclear weapons complex' and hosts the National Nuclear Security Administration—is the 'most privatized federal department', with 94% of its expenditure going to a 'handful' of contractors in 2004.[167] This situation leads these authors to conclude that 'it is now hard to tell where government ends and where the corporations which comprise and profit from its activities begin', describing these corporations as 'parastatal' but, crucially, not subject to any sort of democratic accountability.[168]

Indeed, the general lack of accountability and transparency surrounding the US's nuclear weapons system makes it very difficult to ascertain its true cost. For example, in 2005 the Government Accountability Office reported that even the DOD itself did not know the exact cost of the US nuclear arsenal.[169] Estimates on the cost of the US's nuclear weapons therefore vary, although Stephen Schwartz has estimated that the US spent $8.7 trillion on its nuclear forces between 1940 and 2010.[170] The opaque nature of the system is thus clearly a significant barrier to democratic deliberation and participation in decision-making concerning nuclear matters and something that needs to be rectified if there is to be democratic control of nuclear weapons pursuant to their elimination.

Critics such as Chomsky thus emphasise that the essential task of tackling the Pentagon System involves recognising that the problem is one of 'power and privilege' and 'specific institutional structures'.[171] The scale of this challenge requires the construction of 'stable popular institutions' so that citizens can act to undermine the power of elites and participate fully in decision-making.[172] Elaine Scarry reaches similar conclusions in her 2014 work *Thermonuclear Monarchy: Choosing between Democracy and Doom,* where she argues that the US people must use the US Constitution as a tool to dismantle the US nuclear weapons system. For Scarry, in a similar fashion to Dahl and Deudney, the possession of nuclear weapons has converted the US government into 'a monarchic form of rule that places all defense in the executive branch of government', leaving the population 'incapacitated'.[173] She goes on to argue that this dire situation is 'radically incompatible' with the US Constitution, first because that document requires a Congressional declaration of war and secondly, because of the 'constitutional requirement that distributes to the entire adult population shared responsibility for use of the country's arsenal'.[174] In response to this problem, Lindsay argues for 'decentralizing authority' because Congress 'should play an active role on nuclear weapons matters regardless of who occupies the White House'.[175] Similarly, a recent report by the Ploughshares Fund outlines how the US can reduce its nuclear spending, reform its nuclear posture and restrain its nuclear war plans so that the nuclear button is controlled by Congress.[176]

Concerning political support for nuclear arms control and disarmament today, Lewis Dunn argues that 'there is now a fragile center-right political consensus for modernizing the U.S. strategic force posture and nuclear infrastructure. But that consensus has opponents in Congress, the think-tank community, and the broader American public.'[177] However, there is also only a small group of Congresspeople who publicly support US action on nuclear disarmament and a TPNW.[178] For example, there are currently seven US members of the international group Parliamentarians for Nuclear Non-proliferation and Disarmament, all of whom are Democrats.[179] Reviewing the positions of Democratic Presidential candidates for the 2020 election, research by the Outrider Foundation found that several of the most prominent candidates supported a No First Use policy, the extension of New START and the ratification of the CTBT, whilst opposing the development of new 'low-yield' nuclear weapons.[180]

However, whilst the Democratic party may appear to have more progressive policies on nuclear weapons than Republicans, when in office the former does not always choose, or is unable, to put these policies into practice. For example, Kristensen has compared President Obama's record with that of the previous presidents holding office during the nuclear age. He found that Obama cut fewer warheads—in terms of numbers rather than percentages—than 'any administration ever' and that 'the biggest nuclear disarmers' in recent decades have been Republicans, not Democrats. Kristensen thus drily observes of this situation that 'a conservative Congress does not complain when Republican presidents reduce the stockpile, only when Democratic presidents try to do so'.[181]

In terms of domestic forces that might attenuate the US's march to nuclear weapons modernisation and help implement progressive reforms, Acheson, Gandenberger, Lichterman and Mello argue that, at present, the 'main obstacle' is likely to be US military industries' seeming inability to complete 'ever-more complex manufacturing and industrial projects'.[182] This is principally because of the 'cost overruns and schedule delays' which have blighted such programmes, 'eroding congressional and military support' and causing this organisation 'to downscale or indefinitely defer' several projects. By contrast, domestic opposition to US nuclear weapons policies is very weak, with 'little debate' amongst the public on this issue and no real 'disarmament movement' to speak of.[183] Similarly, Falk and Krieger have spoken of the 'general complacency amongst the public', which, Falk states, 'may be hiding an underlying despair, a turning away because it seems impossible to get rid of the weaponry'.[184] Furthermore, Acheson and her co-authors argue that public discourse on nuclear weapons is 'dominated by specialists' who focus on proliferation to NNWS or non-state actors rather than on the risks posed by 'nuclear weapons held as central elements of national security policies in the hands of the world's most powerful states'.[185]

The pessimism displayed by several prominent supporters of US nuclear disarmament regarding the potential role of the public and the relative strength of their movement is of no small significance. First, as previously discussed in Chapter 2, such 'movement pessimism' may be expected from elites who generally malign, ignore or underestimate the potential of popular movements, but not necessarily from within the movements themselves.[186] Secondly, recent polling data and surveys show that, despite the impediments to understanding the issues at a substantive level, the US public supports a range of progressive action. This includes adopting a more restrained and multilateralist approach to global affairs, reducing spending on the military—and nuclear weapons in particular—as well as nuclear arms control, non-proliferation and disarmament. Thus, for Peter Cary of the Center for Public Integrity, on the topic of military spending, 'public opinion is again at odds with Washington'.[187]

Other recent polls from the Pew Research Center and Survey Sampling International found that the majority of Americans want the US to 'focus on its own problems rather than expanding the military's role abroad'.[188] Similarly, when it comes to nuclear weapons, a 2012 Stimson Center study found that two-thirds of those polled 'decreased the budget for nuclear weapons, including eight in ten Democrats and two thirds of Republicans, with the sample as a whole cutting it an average of 27%—the largest area percentage cut'.[189] This data reinforces the points made in the above discussion of *institutional democratisation*, highlighting the gap between US public opinion on military and nuclear matters and government policy. In terms of how this gap might be closed, one of the major difficulties in the US shifting to a new approach will be how to reconcile the sometimes contradictory positions taken by the public on these issues. For example, according to Andrew Kohut of Pew Research, 'the typical American continues to look at world leadership with a fair degree of skepticism and is extremely wary of engagement in areas of conflict. At the same time, most continue to take considerable comfort in American military power.'[190]

It is worth considering at this point the difference between the US disarmament movement of today and that of its peak in the 1980s, which saw in New York in 1982 the largest protest, at that time, in US history, with the theme of 'Freeze the Arms Race—Fund Human Needs'.[191] For one veteran disarmament activist I spoke to, the Freeze campaign was part of a 'broader and deeper' social movement than exists today, proposing disarmament but also critical of nuclear power and with strong analyses of political institutions.[192] Yet the Freeze movement has been criticised by other campaigners for not being an abolition movement, but based around 'removing the sense of fear' regarding the US–Soviet arms race.[193] Meyer's *A Winter of Discontent*, which tells the story of the Freeze movement and analyses the politics of peace movements, also notes the way in which mainstream politicians co-opted and 'demobilized' the campaign, turning it into something 'more moderate and less threatening to the bipartisan tradition of arms control'.[194]

In their discussion of the 'nuclear freeze' movement, Robert Entman and Andrew Rojecki also describe the nature of the public discourse regarding nuclear weapons issues in the US, whereby 'news media maintain public quiescence and even support in the face of a government policy that opposes stable and perhaps deeply felt majority opinion—opinion actively voiced by a large, organized, and determinedly mainstream political movement'.[195] This raises important questions, such as whether the information compiled by the many highly knowledgeable analysts of the US's nuclear policies, will reach, in a cogent form, the American public so that political participation can grow and be based on sound information. Furthermore, if nuclear disarmament is to have a high profile in the US, it is likely that mainstream and alternative media will need to provide platforms for a sustained debate covering a wide range of policy options, going beyond elite preferences alone.

In the medium to long term, the key challenge for opponents of the US's nuclear weapon system is thus how to engage with and mobilise the public in order to lever out undemocratic and militarist forces from systems of governance and gain popular control over key institutions. For example, in order for such movements to be built it is likely that there will need to be a common understanding amongst politically active groups that nuclear disarmament requires the MIC or Pentagon System to be dismantled as part of wider progressive social, economic and political change. Moreover, whilst some degree of elite bargaining and coalition building may be necessary to develop legislation and highlight key issues, history suggests that disarmament activists need to be wary of politicians' personal, narrow agendas if truly progressive policies are to be developed.

In addition to looking at the nature of US nuclear weapons policy and decision-making, it is useful to contextualise such processes within the overall US political system, for example, to compare how the conduct of nuclear politics relates to wider political cultures and trends—including the present health of US democracy. In doing so, we may shed some light on the relationship between nuclear weapons decision-making and *institutional democratisation* and the obstacles to and opportunities for the latter in the US today.

Indeed, it is reasonable to suggest initially that the nature of US nuclear weapons decision-making outlined above is just one example of the far-reaching influence of small groups of extremely powerful and wealthy individuals and corporations over the US political system. Such concerns have led some scholars to question whether the US today can accurately be described as a functioning democracy. For example, Brenner claims the US is a 'plutocracy', whilst the results of Gilens and Page's investigation into which actors exert most influence over US public policy found 'substantial support for theories of Economic Elite Domination'.[196] Elsewhere Sheldon Wolin describes the emerging political system as 'inverted totalitarianism' given the increasing power of an authoritarian state run by and for the rich and the decline of institutions capable of checking that power, whilst Markovits and Ayres argue that the 'distortions' of the US's 'election process' mean that it is subject to 'perpetual' and 'undemocratic minority rule' favouring the Republican party.[197]

When reviewing the publications of mainstream research groups, it is notable that they have quite a different view of the state of US democracy than the one presented above. For example, the Economist Intelligence Unit, from 2006 to 2015, classified the US as a 'full democracy', before it was judged to have slipped to the rank of 'flawed democracy' in 2016.[198] Elsewhere, a 2017 Freedom House report gave the US a score of 89/100 and classified it as in the first out of seven ranks for political rights and civil liberties.[199] Such discrepancies between critical and mainstream approaches should lead us to question the methodologies used by both sides to reach their results. For example, from an initial review of this literature it is not apparent that the gap between public opinion and government policy on nuclear matters, as outlined above, is taken into account when formulating assessments of the health of US democracy. Chapter 8 will provide a fuller consideration of how such methodological discrepancies might be overcome by assessing the ways in which nuclear possession's impact on democratic processes might be measured.

## 3.6  Conclusion

In terms of NWS nuclear disarmament, the preceding discussion has shown that we must recognise that the US sits in a category by itself given both the size and scale of its nuclear arsenal and its overall military power, which together exert a singular influence on all other nation's strategic thought and nuclear choices. Moreover, US nuclear possession was, for most of the Cold War, justified in terms of the need to defend democracy in the ideological struggle against communism. For these reasons, it is clear that the main contention of this study—that *institutional democratisation*: i) has significant analytical value given the limitations of existing approaches to the politics of nuclear possession and disarmament, and ii) is necessary, on a practical level as a transformative political process, for NWS nuclear disarmament—will, to a great extent, stand or fall depending upon its applicability to the US case.

With regard to the former claim, this chapter found that the US nuclear weapons establishment is today embedded, at the domestic level, in a range of governmental and non-governmental economic, political and social institutions that have an interest in the bomb's continual reproduction. This interest has grown to be far deeper and more widespread than when it was originally formed, during the turbulent economic, social and political conditions of WW2. The security model has most validity at this historical point, as the drive for the bomb emerged at a time when all of US society was geared towards defeating Nazi Germany.

The secretive and hierarchical nature of the bomb thus reflected the regimented nature of the US polity during this period, so that the US government was able to justify nuclear possession and the atomic bombing of Japan with reference to the onerous demands of war and the imperative need for victory. Yet, as the critical literature explains, rather than the US military establishment being downgraded after the war and nuclear possession being a temporary phenomenon—so that both became subject to democratic control and restraint—the Pentagon System and national security state grew in power whilst efforts at international management of the bomb failed. With such missed opportunities the nuclear revolution thus took on ominous new domestic and international meanings with the emergence of the Cold War.

For example, *institutional democratisation* explains how the entrenchment of the US's nuclear status had significant consequences for all aspects of US domestic political life. As Dahl noted, the US bomb, justified officially as an essential tool to defend democracy, represented, from the beginning, a 'tragic paradox'. This was because the production and growth of the nuclear weapons complex necessitated the suspension of democracy in important areas of government and the creation of a 'guardianship', or, for Deudney and Scarry, 'monarchy'. This situation became embedded over the course of the 20th century because the Cold War was framed at an official level as an ideological conflict involving the US defending its liberal and benign democracy against Soviet communism's evil empire. Governments on both sides thus used the Cold War as a means of managing their populations and legitimising the maintenance and growth of enormous military power, including their nuclear arsenals. Moreover, as the US's military and nuclear might grew in line with its global ambitions, so did its influence and importance to the US economy and society. This made it particularly difficult for critics to challenge possession of the bomb, not least because defenders of the status quo could accuse their opponents of siding with the ideological enemy—a particularly potent means of protecting their own domestic interests.

However, nuclear technology, whilst ensuring significant levels of unaccountable power for small groups of domestic elites, soon came to threaten the very survival of the US as it spread to the Soviet Union, and beyond. Moreover, nuclear possession was and is important to the US for ensuring world order is run in accordance with its interests, including via power projection and extended deterrence in Europe and East Asia. Yet neither of these points are considered by the security model, which principally frames nuclear acquisition as a necessary and rational response to external threats to sovereignty and national security. Critical approaches such as *institutional democratisation* thus better explain why the US's nuclear weapons complex not only survived but was modernised following the end of the Cold War and the disappearance of the US's primary strategic opponent. Other limitations of the mainstream and realist literature were shown, as the potentially significant, but hitherto unfulfilled, domestic sources of international cooperation on nuclear arms control and disarmament action to be found amongst the US public were identified.

When it comes to bilateral and multilateral disarmament processes, the US has a particular responsibility to boost cooperation in relation to nuclear disarmament action in numerous ways, beginning with strategic restraint involving the many new and varied types of emerging military technology and the threat and use of force. However, in the post-Cold War era successive US administrations, despite much rhetoric about the desirability of a NWFW, have failed to live up to these responsibilities. Instead, democratic processes have been disregarded, with decisions regarding nuclear weapons use concentrated in the hands of the President, significant support given to allies' nuclear weapons programmes and the will of the US public on nuclear weapons issues largely ignored. International law and the UN Charter, meanwhile, have repeatedly been sidelined in favour of military build-ups and overseas aggression.

At home huge state intervention in the US economy through massive military spending—much of which goes into private hands—is presented as being necessary for the government to protect the nation from dangerous overseas threats. However, the US government's behaviour, at home and abroad, consciously increases threats to its own national security. Whilst decisionmakers are aware of the consequences of their policies, they prefer to discuss how to counter the threats they've helped create rather than considering alternatives to militarism. Such alternatives and the causes of insecurity are therefore largely absent from the mainstream discourse, which instead tends to repeat the narratives of the powerful. The result is a self-fulfilling prophecy as defenders of the status quo and an even more powerful military, use the spread of WMD and the threat of terrorism to generate an atmosphere of fear and uncertainty in order to justify expansionist policies.

Taken as a whole, the US's government's policies have thus severely impeded progress on nuclear non-proliferation and disarmament, both because its quest for hegemony and full spectrum dominance provokes other nations into seeking or retaining the bomb and because the US has greatly accelerated the spread of nuclear weapons internationally by supporting the nuclear programmes of its allies. In doing so, the US has substantially increased the possibility that someday someone will use nuclear weapons, whether by accident, miscalculation, madness or design. Given the significant degree of continuity exhibited by the US's behaviour post-1945, China and Russia understand the US's strategic intent and therefore prioritise defensive measures against a potential US attack, within which nuclear deterrence plays a central role.

Consequently, there is a particularly long list of measures the US could unilaterally take to begin fulfilling its obligations under the NPT, in order to both eliminate its own nuclear arsenal and create the conditions whereby other NWS are able to do so. These unilateral actions should be implemented first as progress on them would make it much easier to take forward other cooperative bilateral and/or multilateral measures between NWS. In the medium to long term, efforts both to eliminate the US's nuclear weapons and transform US foreign policy in line with domestic and international law require challenging the deep rooted institutions and interests of the MIC/ Pentagon System. However, supporters of the permanent war economy have not only successfully resisted efforts to reduce its role in society but have ensured its growth in recent years, so that US military spending now dwarfs that of the Reagan administration at the height of the Cold War, whilst the cost of the US's nuclear weapons system is expected to grow substantially over the next 20 years as the triad is modernised.

These developments are partly due to the relative weakness of current oppositional forces, in particular the nuclear disarmament movement in the

US and globally. A key obstacle to building such movements is the secrecy and misinformation that surrounds nuclear weapons. For example, despite the openness of the US's formally democratic institutions, the vast majority of the US public is not aware of how their leaders use nuclear weapons, so that, whilst there is significant opposition to current policies—such as first use—this does not translate into political pressure for change. Moreover, research shows that the US public also generally supports progressive action towards nuclear disarmament and has historically succeeded in pressuring Washington into arms control agreements. In addition, there has been a lively debate within elite circles regarding the value of nuclear possession, with prominent figures such as William Perry and General George Lee Butler advocating nuclear abolition. Despite these positive factors, there exists a sense of 'movement pessimism' or scepticism in the US nuclear disarmament movement about the potential role the public could play in these areas, which may prevent popular forces being used effectively to realise the movement's goals.

In terms of opportunities for nuclear disarmament action then, harnessing public opinion is key if decision-makers are to be held accountable to domestic and international law and Congress and the President are to be empowered to take progressive action on these issues. Whilst the *guardianship* model of arms control and disarmament has had an important impact, in terms of providing some boundaries and stability to the US–Russian nuclear competition, it is also clearly limited in terms of how far it can take the project of eliminating nuclear weapons. In terms of moving forward, in the short term, the *guardianship* approach could be beneficial to disarmament advocates, for example, if opportunities are taken to collaborate with actors from within the US establishment who are critical of nuclear possession. In the longer term, if a democratic peace is indeed to be realised, social movements must find means to organise effectively in order to democratise the US polity. This transition to a deeper democracy at home and an international policy based on diplomacy, cooperation and multilateralism, will be necessary for Washington to undertake if it is to persuade other nations of its peaceful intentions and mitigate their fear and uncertainty over the long term, particularly given the US's singular military might and record of power projection.

## Notes

1  Rhodes, Richard (1986) *The Making of the Atomic Bomb* (New York: Simon & Schuster), 379.
2  US Department of Energy (2017) *Negotiating International Control*, www.osti.gov.
3  Nichols, Tom and Struckman, Dana (2017) Welcome to American Nuclear Weapons 101, http://nationalinterest.org, 3 October.
4  Spalding, Robert (2013) Nuclear Weapons Are the U.S.'s Instruments of Peace, www.washingtonpost.com, 4 October.
5  Ibid.
6  Kaufman, Joyce P. (2013) *A Concise History of U.S. Foreign Policy*, 3rd Edition (Lanham, MD: Rowman & Littlefield), 78.

7  Moore, Mike (1996) Behind the Clock Move, *Bulletin of the Atomic Scientists*, 52(2), March/April, 22.

8  Ibid.

9  Pringle, Peter and Arkin, William (1984) *S.I.O.P: The Secret U.S. Plan for Nuclear War* (New York: W. W. Norton), 245.

10 Hey, Nigel (2000) Survey Shows Public Concerned over National Security, Still Supports Nuclear Arsenal, www.nuclearfiles.org, 25 August.

11 Sagan, Scott and Valentino, Benjamin (2017) Revisiting Hiroshima in Iran: What Americans Really Think about Using Nuclear Weapons and Killing Noncombatants, *International Security*, 42(1), Summer, 5, 79.

12 Perry, William J. (2015) *My Journey at the Nuclear Brink* (Stanford, CA: Stanford University Press), xvi, 195.

13 US Department of Defense (2010) *Nuclear Posture Review Report*, www.defense. gov, April, i.

14 US Department of Defense (2014) *Quadrennial Defense Review*, www.defense. gov, 13; Kristensen, Hans M., Norris, Robert S. and Oelrich, Ivan (2009) *From Counterforce to Minimal Deterrence: A New Nuclear Policy on the Path Toward Eliminating Nuclear Weapons* (Washington, DC: Federation of American Scientists), 8–9; US Joint Chiefs of Staff (2015) *National Military Strategy*, www. jcs.mil, 11.

15 Obama, Barack (2009) Remarks by President Barack Obama in Prague, https:// obamawhitehouse.archives.gov, 5 April.

16 US Department of Defense (2014) *Quadrennial Defense Review*, www.defense. gov, 13.

17 Ford, Christopher (2007) *Cluster 1: Disarmament*, 2007 Preparatory Committee Meeting of the Treaty on the Non-Proliferation of Nuclear Weapons, 8 May, Vienna, www.nti.org, 6.

18 Brown, Harold and Deutch, John (2007) The Nuclear Disarmament Fantasy, *Wall Street Journal,* 19 November.

19 Ibid.; Hoffman, David E. (2014) Review: 'Reagan at Reykjavik Forty-Eight Hours that Ended the Cold War' by Ken Adelman, www.washingtonpost.com, 9 May; US Office of the Historian (2017) *The Acheson-Lilienthal & Baruch Plans, 1946*, https://history.state.gov; The Reagan Vision (2017) The Reykjavik Summit, www. thereaganvision.org.

20 US Joint Chiefs of Staff, *National Military Strategy*, 2.

21 Hudson, John and Sonne, Paul (2020) Trump Administration Discussed Conducting First U.S. Nuclear Test in Decades, www.washingtonpost.com, 23 May.

22 Chalfant, Morgan (2015), Top Pentagon Generals Deem Russia Largest 'Existential Threat' to U.S., *Washington Free Beacon*, https://freebeacon.com, 15 July.

23 Lambakis, Steven (2018) Foreign Space Capabilities: Implications for U.S. National Security, *Comparative Strategy*, 37(2), 100.

24 Hudson and Sonne, Trump Administration Discussed.

25 Panda, Ankit and Narang, Vipin (2019), The Hanoi Summit Was Doomed from the Start, www.foreignaffairs.com, 5 March.

26 CSIS (2020), *Missiles of North Korea*, https://missilethreat.csis.org, 4 March.

27 Acton, James M. (2013) *Silver Bullet* (Washington DC: Carnegie Endowment for International Peace), 15.

28 US Department of Defense (2018) *Nuclear Posture Review*, https://dod.defense. gov, 33.

29 Borger, Julian (2020) Deployment of New US Nuclear Warhead on Submarine a Dangerous Step, Critics Say, www.theguardian.com, 29 January.

30 Arkin, William (2020) Risk of Nuclear War Rises as U.S. Deploys a New Nuclear Weapon for the First Time since the Cold War, www.democracynow.org, 7 February.

31 *New York Times* (2016) Donald Trump Expounds on his Foreign Policy Views, www.nytimes.com, 26 March; McKenzie, Pete (2020) America's Allies Are Becoming a Nuclear-Proliferation Threat, www.defenseone.com, 25 March.

32 Roberts, Brad (2013) *Extended Deterrence and Strategic Stability in Northeast Asia*, www.nids.go.jp, 23.

33 Chomsky, Noam (1992) *Deterring Democracy* (London: Vintage); Melman, Seymour (1976) *The Permanent War Economy* (London: Simon & Schuster); Kolko, Joyce and Kolko, Gabriel (1972) *The Limits of Power: The World and United States Foreign Policy, 1945–1954* (New York: Harper & Row); Kaldor, Mary (1990) *The Imaginary War: Understanding the East–West Conflict* (Oxford: Blackwell).

34 Chomsky, *Deterring Democracy*, 21.

35 Jablonsky, David (2002) The State of the National Security State, *Parameters*, Winter, 5.

36 Kaufman, *Concise History of U.S. Foreign Policy*, 78.

37 McFarland, Linda (2001) *Cold War Strategist: Stuart Symington and the Search for National Security* (Santa Barbara, CA: Praeger), 113.

38 STRATCOM (1995) Essentials of Post-Cold War Deterrence, www.nukestrat. com.

39 Kristensen, Hans M. (2012) *Nuclear Studies and Republican Disarmers*, https://fas. org, 16 February.

40 Chomsky, Noam (2003) Dominance and its Dilemmas, *Boston Review,* October; Leffler, Melvyn (1993) *Preponderance of Power: National Security, Truman Administration and the Cold War* (Redwood City, CA: Stanford University Press).

41 Lake, Anthony (1993) From Containment to Enlargement, http://fas.org/news/ usa/1993/usa-930921.htm.

42 Rhodes, *Making of the Atomic Bomb*, 379; Gosling, F. C. (2010) *The Manhattan Project: Making the Atomic Bomb* (Washington, DC: US Department of Energy), 97.

43 Atomic Heritage Foundation (2013) *Transforming the Relationship between Science and Society: The Manhattan Project and its Legacy*, http://www.atomicheritage. org, 14–15 February, 11.

44 Wills, Gary (2010) *Bomb Power: The Modern Presidency and the National Security State* (London: Penguin), 1.

45 Eisenhower, Dwight D. (1961), Farewell Address, www.ourdocuments.gov.

46 Dahl, Robert A. (1985) *Controlling Nuclear Weapons: Democracy versus Guardianship* (Syracuse, NY: Syracuse University Press), 3.

47 Meyer, David S. (1990) *A Winter of Discontent: The Nuclear Freeze and American Politics* (Santa Barbara, CA: Praeger), 27.

48 Ellsberg, Daniel (2011) Roots of the Upcoming Nuclear Crisis, in Krieger, David, ed., *The Challenge of Abolishing Nuclear Weapons,* (New Brunswick, NJ: Transaction), 54.

49 Meyer, *Winter of Discontent*, 27.

50  Nolan, Janne E. (1989) *Guardians of the Arsenal: The Politics of Nuclear Strategy* (New York: Basic Books), 69; Nolan, Janne E. (1999) *An Elusive Consensus: Nuclear Weapons and American Security After the Cold War* (Washington, DC: Brookings); Carter, April (1989) *Success and Failure in Arms Control Negotiations* (Oxford: Oxford University Press), 25.

51  Lindsay, James M. (1991) *Congress and Nuclear Weapons* (Baltimore, MD: Johns Hopkins); Pringle and Arkin, *S.I.O.P.*

52  Sauer, Tom (2005) *Nuclear Inertia: US Nuclear Weapons Policy After the Cold War* (London: I. B. Tauris).

53  Talbott, Strobe (1984) *Deadly Gambits: The Reagan Administration and the Stalemate in Nuclear Arms Control* (London: Pan), 229.

54  Sauer, *Nuclear Inertia*, 25.

55  Kristensen et al., *From Counterforce to Minimal Deterrence*, 1.

56  Sauer, *Nuclear Inertia*, 57.

57  Cabasso, Jacqueline (2008) Redefining Security in Human Terms, in Spies, Michael and Burroughs, John, eds., *Nuclear Disorder or Cooperative Security? US Weapons on Terror, the Global Proliferation Crisis, and Paths to Peace*, http://wmdreport. org, 33.

58  Sauer, *Nuclear Inertia*, 83.

59  Ibid., 83, 94.

60  Ibid., 167.

61  Nolan, *Elusive Consensus*.

62  Cortright, David (2008) *Peace: A History of Movements and Ideas* (Cambridge: Cambridge University Press), 149–150, 323.

63  Knopf, Jeffrey W. (1998) *Domestic Society and International Cooperation: The Impact of Protest on US Arms Control Policy* (Cambridge: Cambridge University Press), 247.

64  Meyer, *A Winter of Discontent*.

65  Kull, Steven (2004) *Americans on WMD Proliferation* (College Park, MD: University of Maryland), 9.

66  US in the World (2009) *Talking about Nuclear Weapons with the Persuadable Middle* (Washington DC: US in the World Initiative), 14.

67  Ford, Christopher (2007) Cluster 1: Disarmament, *2007 Preparatory Committee Meeting of the Treaty on the Non-Proliferation of Nuclear Weapons*, 8 May, Vienna, www.nti.org, 1.

68  van der Zeijden, Wilbert, Snyder, Suzi and Ekker, Peter Paul (2012) *Exit Strategies: The Case for Redefining NATO Consensus on NATO T.N.W.*, http:// nonukes.nl, 8.

69  Kristensen et al., *From Counterforce to Minimal Deterrence*, 2.

70  Steinbrunner, John and Gallagher, Mary (2008) If you Lead, they Will Follow: Public Opinion and Repairing the U.S.-Russian Strategic Relationship, www.armscontrol.org, January/February.

71  Alan J. Kuperman (2015), *Obama's Libya Debacle: How a Well-Meaning Intervention Ended in Failure*, www.foreignaffairs.com. March/April.

72  Una Becker, Harald Müller and Simone Wisotzki (2008), Democracy and Nuclear Arms Control—Destiny or Ambiguity? *Security Studies*, 17(4), 848.

73  Ibid.

74  Ibid., 825.

75 Ritchie, Nick (2008) *US Nuclear Weapons Policy After the Cold War: Russians, 'Rogues' and Domestic Division: The Evolution of Strategy and Policy* (London: Routledge), 6–10.

76 Lindsay, James M. (1991) *Congress and Nuclear Weapons* (Baltimore, MD: Johns Hopkins), 163.

77 McLean, *Nuclear Weapons Decisions*, 256.

78 Ibid., 62.

79 Ibid., 63.

80 Gerson, Joseph (2007) *Empire and the Bomb* (London: Pluto), 20.

81 Ibid., 215–216.

82 Burr, William (2008) *'Prevent the Reemergence of a New Rival': The Making of the Cheney Regional Defense Strategy, 1991–1992*, http://nsarchive.gwu.edu

83 Ibid.

84 Arkin, William (1993) Nuclear Junkies: Those Lovable Little Bombs, *Bulletin of the Atomic Scientists*, July/August, 24.

85 Ritchie, *US Nuclear Weapons Policy*, 32.

86 Arkin, Nuclear Junkies, 22–27.

87 Ibid.

88 Sahay, Usha and Reif, Kingston (2013) *Non-Treaty Cuts to the U.S. Nuclear Stockpile* (Washington, DC: Center for Arms Control and Non-Proliferation), 7.

89 Ritchie, *US Nuclear Weapons Policy*, 22.

90 Kristensen, Hans M. (1998) *Proliferation of Weapons of Mass Destruction and US Nuclear Strategy* (London: BASIC), 33.

91 Nolan, *Elusive Consensus*; Schell, Jonathan (1998) *The Gift of Time* (London, Granta); Falk, Richard and Krieger, David (2008) *At the Nuclear Precipice: Catastrophe or Transformation?* (London: Palgrave Macmillan), 34.

92 Arkin, Nuclear Junkies; Sauer, *Nuclear Inertia*, 64.

93 Sauer, *Nuclear Inertia*, 152.

94 Kristensen, Hans (2003) *Changing Targets II: A Chronology of U.S. Nuclear Policy Against Weapons of Mass Destruction* (London: Greenpeace), 4.

95 Leffler, Melvyn P. (2009) Think Again: Bush's Foreign Policy, www.foreignpolicy.com, 23 October.

96 Burr, *'Prevent'*.

97 Mann, James (2004) The True Rationale? It's a Decade Old, www.washingtonpost.com, 7 March.

98 US Joint Chiefs of Staff (2000) *Joint Vision 2020* (Washington, DC: US Government Printing Office), 1.

99 Ikenberry, John G. (2002) America's Imperial Ambition, *Foreign Affairs*, September–October.

100 Goodby, James E. (2006) *At the Borderline of Armageddon: How American Presidents Managed the Atom Bomb* (Lanham, MD: Rowman & Littlefield), 188–189.

101 Boese, Wade (2002) U.S. Withdraws from ABM Treaty; Global Response Muted, www.armscontrol.org, July/August.

102 Korb, Lawrence J., Conley, Laura and Rothman, Alex (2011) A Historical Perspective on Defense Budgets, www.americanprogress.org, 6 July.

103 Lyons, Marco J. (2014) *The Role of Nuclear Weapons in National Security: The 2010 Nuclear Posture Review*, https://medium.com, 14 July.

104    Obama, Remarks in Prague; The Nobel Prize (2009), The Nobel Peace Prize for 2009, www.nobelprize.org, 9 October.

105    Cirincione, Joseph (2013) *Nuclear Nightmares: Securing the World Before it Is Too Late* (New York: Columbia University Press), 34.

106    Priest, Dana (2012) U.S. Nuclear Arsenal Is Ready for Overhaul, www.washingtonpost.com, 15 September.

107    Congressional Budget Office (2013) *Projected Costs of U.S. Nuclear Forces, 2014 to 2023,* https://www.cbo.gov, 20 December, 2; Wolfstahl, J. B, Lewis, J. and Quint, M. (2014) *The One Trillion Dollar Triad: US Strategic Nuclear Modernization over the Next Thirty Years* (Monterey, CA: James Martin Center for Nonproliferation Studies), 4; Reif, Kingston (2018) U.S. Nuclear Modernization Programs, www.armscontrol.org, August.

108    Lichterman, Andrew (2010) The START Treaty and Disarmament: A Dilemma in Search of a Debate, http://www.wslfweb.org, December 6.

109    Izbicki, Michael (2010) What's Wrong with America's Nuclear Hawks? *Strategic Studies Quarterly*, Winter, 138.

110    Zala, Ben and Futter, Andrew (2013) *A Sustainable Approach to Nuclear Zero: Breaking the Nuclear–Conventional Link* (London: Oxford Research Group), 2–3.

111    Zagorski, Andrei (2011) *Russia's Tactical Nuclear Weapons: Posture, Politics and Arms Control* (Hamburg: IFSH); Acton, James M. (2013) *Silver Bullet* (Washington, DC: Carnegie Endowment for International Peace); Hansell, Cristina and Potter, William C., eds. (2009) *Engaging China and Russia on Nuclear Disarmament* (Monterey, CA: James Martin Center for Nonproliferation Studies), 2.

112    Perkovich, George and Acton, James M., eds. (2009) *Abolishing Nuclear Weapons: A Debate* (Washington, DC: Carnegie Endowment for International Peace), 30–31.

113    Lonsdale, David J. (2019) The 2018 Nuclear Posture Review: A Return to Nuclear Warfighting? *Comparative Strategy*, 38(2), 103.

114    US Department of Defense, *Nuclear Posture Review*, 21.

115    Work, Robert (2015) *The Third U.S. Offset Strategy and its Implications for Partners and Allies*, www.defense.gov, 28 January.

116    Chin, Warren (2019) Technology, War and the State: Past, Present and Future, *International Affairs,* 95(4), 774.

117    President of the United States (2017) *National Security Strategy*, www.whitehouse.gov, 25.

118    US Department of Defense (2018) *National Defence Strategy*, https://dod.defense.gov, 4.

119    Simón, Luis (2016) The 'Third' US Offset Strategy and Europe's 'Anti-Access' Challenge, *Journal of Strategic Studies*, 39(3), 417.

120    Ibid.

121    Woolf, Amy (2019) *Conventional Prompt Global Strike and Long-Range Ballistic Missiles: Background and Issues* (Washington, DC: Congressional Research Service), 3.

122    Ibid., 2.

123    Wilkening, Dean (2019) Hypersonic Weapons and Strategic Stability, *Survival*, 61(5), 129.

124    Ibid.

125    Ibid., 136.

126 Ibid., 138.
127 Ibid.
128 Futter, Andrew (2018) *Hacking the Bomb: Cyber Threats and Nuclear Weapons* (Washington, DC: Georgetown University Press), 4.
129 Cimbala, Stephen J. (2014) Cyber War and Deterrence Stability: Post-START Nuclear Arms Control, *Comparative Strategy*, 33(3), 279.
130 Gartzke, Erik and Lindsay, Jon R. (2017) Thermonuclear Cyberwar, *Journal of Cybersecurity*, 3(1), 38.
131 Acton, James M. (2018) Escalation through Entanglement: How the Vulnerability of Command-and-Control Systems Raises the Risks of an Inadvertent Nuclear War, *International Security*, 43(1), Summer, 56.
132 Ibid., 58–60.
133 Payne, Kenneth (2018) Artificial Intelligence: A Revolution in Strategic Affairs? *Survival*, 60(5), 26.
134 Ibid., 24.
135 Fatton, Lionel P. (2016) The Impotence of Conventional Arms Control: Why Do International Regimes Fail When they Are Most Needed? *Contemporary Security Policy*, 37(2), 203.
136 Brown and Deutch, Nuclear Disarmament Fantasy.
137 Arbatov, Alexei (2015) *An Unnoticed Crisis: The End of History for Nuclear Arms Control?* https://carnegieendowment.org, June.
138 Pifer, Steven (2016) New START Turns Five, www.brookings.edu, 4 February.
139 Arbatov, Alexei (2019) Mad Momentum Redux? The Rise and Fall of Nuclear Arms Control, *Survival*, 61(3), 34.
140 Arbatov, Alexei (2017) Understanding the US–Russia Nuclear Schism, *Survival*, 59(2), 41.
141 Ibid.
142 Ibid., 17.
143 Ibid., 61.
144 Arbatov, Alexei (2019) Mad Momentum Redux? The Rise and Fall of Nuclear Arms Control, *Survival*, 61(3), 27.
145 Chin, Technology, War and the State, 95(4), 783.
146 Talmadge, Caitlin (2019) Emerging Technology and Intra-War Escalation Risks: Evidence from the Cold War, Implications for Today, *Journal of Strategic Studies*, 42(6), 864.
147 Ibid., 883.
148 Pollack, Joshua H. (2015) Boost-Glide Weapons and US-China Strategic Stability, *Nonproliferation Review*, 22(2), 155.
149 Arbatov, Understanding the US–Russia Nuclear Schism, 53.
150 Scheber, Thomas and Guthe, Kurt (2013) Conventional Prompt Global Strike: A Fresh Perspective, *Comparative Strategy*, 32(1), 25.
151 Wilkening, Dean (2019) Hypersonic Weapons and Strategic Stability, *Survival*, 61(5), 144.
152 Futter, Andrew (2016) *Cyber Threats and Nuclear Weapons: New Questions for Command and Control, Security and Strategy* (London: RUSI Occasional Paper, July), 1–2.
153 Ibid., 39.
154 Marchant, Gary E. and Allenby, Brad (2017) Soft Law: New Tools for Governing Emerging Technologies, *Bulletin of the Atomic Scientists*, 73(2), 112.

155  Ibid.

156  Altmann, Jürgen and Sauer, Frank (2017) Autonomous Weapon Systems and Strategic Stability, *Survival*, 59(5), 133–134.

157  Payne, Kenneth (2018) Artificial Intelligence: A Revolution in Strategic Affairs? *Survival*, 60(5), 19.

158  Article 36 (2019), Critical Commentary on the 'Guiding Principles', www.article36.org, November.

159  Women's International League for Peace and Freedom (2014) Military-Industrial Complex, www.reachingcriticalwill.org.

160  Lichterman, Andrew (2012) United States, in Acheson, Ray, ed., *Assuring Destruction Forever: Nuclear Weapon Modernization Around the World* (New York: Reaching Critical Will), 91.

161  Chomsky, Noam (1989) *Necessary Illusions: Thought Control in Democratic Societies* (London: Pluto Press), 185; Johnson, Chalmers (2008) *Tomgram: Chalmers Johnson, How to Sink America*, www.tomdispatch.com.

162  Lichterman, 'United States', 102.

163  Chomsky, Noam (1993) The Pentagon System, www.thirdworldtraveler.com.

164  Hartung, William D. and Anderson, Christine (2012) *Bombs versus Budgets: Inside the Nuclear Weapons Lobby,* www.ciponline.org, 4.

165  Ibid., 4, 8.

166  Ritchie, *US Nuclear Weapons Policy*, 10.

167  Hill, Damon and Mello, Greg (2006) *Competition or Collusion? Privatization and Crony Capitalism in the Nuclear Weapons Complex: Some Questions from New Mexico*, www.lasg.org, 30 May, 4.

168  Ibid., 10.

169  Government Accountability Office (2005) *Military Transformation: Actions Needed by DOD to More Clearly Identify New Triad Spending and Develop a Long-term Investment Approach*, www.gao.gov, 4 August.

170  Schwartz, Stephen I. (2012) *Unaccountable: Exploring the Lack of Budgetary Transparency for U.S. Nuclear Security Spending*, www.nti.org, 5 January.

171  Chomsky, *Deterring Democracy*, 32.

172  Chomsky, Noam (2002) *Understanding Power: The Indispensable Chomsky* (New York: New Press), 185.

173  Scarry, Elaine (2014) *Thermonuclear Monarchy: Choosing between Democracy and Doom* (New York: Norton), 13.

174  Ibid., 31.

175  Lindsay, James M. (1991) *Congress and Nuclear Weapons* (Baltimore, MD: Johns Hopkins), 170.

176  Benedict, Kennette (2016) Add Democracy to Nuclear Policy, in Collina, Tom Z, and Wilson, Geoff, eds., *10 Big Nuclear Ideas for the Next President* (San Francisco, CA: Ploughshares Fund), 25–27.

177  Dunn, Lewis A. (2016) *Redefining the U.S. Agenda for Nuclear Disarmament* (Livermore, CA: Lawrence Livermore National Laboratory), 39.

178  Nuclear Ban US (2019), Who has Signed ICAN Pledge in US Congress?, www.nuclearban.us, April.

179  PNND (2020) PNND in United States of America, www.pnnd.org.

180  Drozdenko, Tara (2020) Where Do the Democratic Candidates Stand on Nuclear Weapons? https://outrider.org.

181 Kristensen, Hans M. (2014) *How Presidents Arm and Disarm*, https://fas.org, 15 October.

182 Acheson, Ray, Gandenberger, Mia, Lichterman, Andrew and Mello, Greg (2014) United States, in Acheson, Ray, ed., *Still Assuring Destruction Forever: An Update to the 2012 Report: Assuring Destruction Forever: Nuclear Weapon Modernization around the World* (New York: Reaching Critical Will/WILPF), 24.

183 Ibid., 25.

184 Falk, Richard and Krieger, David (2012) *The Path to Zero: Dialogues on Nuclear Dangers* (London: Paradigm), 8.

185 Acheson et al., United States, 25.

186 Nebel, Jacob (2012) The Nuclear Disarmament Movement: Politics, Potential, and Strategy, *Journal of Peace Education*, 9(3), 232.

187 Cary, Peter (2017) The Public Favors Cutting Defense Spending, Not Adding Billions More, New Survey Finds, www.publicintegrity.org, 23 March.

188 Pew Research Center (2017) Globally, More People See U.S. Power and Influence as a Major Threat, www.pewresearch.org, 1 August; Dyer, Geoff (2016) US Voters Wary of Expanding Military Presence Overseas, www.ft.com, 27 October; O'Toole, Molly and De Luce, Dan (2016) The 2016 Election Turned the Politics of Foreign Policy on its Head, http://foreignpolicy.com, 2 November.

189 Kull, Steven, Leatherman, Matthew, Smith, R. Jeffrey (2012) *Consulting the American People on National Defense Spending*, College Park, MD: Program for Public Consultation, University of Maryland, 5.

190 Kohut, Andrew (2014) *Americans: Disengaged, Feeling Less Respected, But Still See U.S. as World's Military Superpower*, www.pewresearch.org, 1 April.

191 Peace Action (2020) Committee for a Sane Nuclear Policy (SANE), www.peaceaction.org.

192 Interviewee: A.

193 Interviewee: C.

194 Meyer, *Winter of Discontent*, 221.

195 Entman, Robert and Rojecki, Andrew (1993) *Silencing the Opposition: Antinuclear Movements and the Media in the Cold War* (Urbana, IL: University of Illinois), 39.

196 Brenner, Michael (2013) Plutocracy in America, www.counterpunch.org, 1 April; Gilen, Martin and Page, Benjamin I. (2014) Testing Theories of American Politics: Elites, Interest Groups and Average Citizens, *Perspectives on Politics*, 12(3), September, 564.

197 Wolin, Sheldon (2010) *Democracy Incorporated: Managed Democracy and the Specter of Inverted Totalitarianism* (Princeton, NJ: Princeton University Press); Markovits, Daniel and Ayres, Ian (2018) The U.S. Is in a State of Perpetual Minority Rule, www.washingtonpost.com, 8 November.

198 Economist Intelligence Unit (2016) *Democracy Index 2016: Revenge of the 'Deplorables'*, ww.eiu.com, 3.

199 Freedom House (2017) *Populists and Autocrats: The Dual Threat to Global Democracy*, https://freedomhouse.org, 24.

# 4 Russia

## 4.1 Introduction

This chapter provides an in-depth investigation of the causes and consequences of Russian nuclear possession and disarmament, in order to further assess the explanatory power of *institutional democratisation* in relation to mainstream and realist approaches. Chapter 8 then contextualises Russia's responsibility for disarmament action alongside the other NWS. This case study proceeds over several sections, beginning with a summary of how the mainstream and realist approaches explored in Chapter 2 relate to Russia's particular experience as an NWS and how Russia has officially justified its nuclear status over time. This is principally done by placing Russia's development of nuclear weapons within the context of relevant historical events from the mid to late 20th and early 21st century—focusing, in particular, on the Cold War.

Section 4.3 begins the process of showing how Russia's nuclear experience illustrates the limitations of the previously reviewed mainstream and realist approaches and the extent to which *institutional democratisation* helps us better understand the causes and consequences of Russian nuclear possession and disarmament. Reviewing alternate perspectives on post-WW2 history in these sections is useful in explaining how international developments interacted with the Soviet Union/Russia's domestic politics and how this interaction led to the rapid expansion of its nuclear weapons establishment. In addition, it is important to consider what opportunities have existed for Russia to adopt an alternative, non-nuclear approach to security—including during the Gorbachev era, the end of the Cold War, as well as today—and imagine how civil society might take advantage of such openings in the future.

Whilst the security model is right to emphasise how the Soviet Union's external threat perceptions drove its original decision to acquire the bomb, we also need to consider the contribution of domestic factors, such as how the militarised and totalitarian nature of the post-WW2 Soviet polity facilitated the rapid development of the bomb and worked to prevent disarmament. Moreover, as the Russian nuclear arsenal grew and developed over the course of the Cold War, the bomb took on important domestic economic, social and political meanings, including as a way by which successive regimes could legitimate their rule.

Today, although Russia's nuclear deterrent strategy is driven by its relations with the US—and, to a lesser extent, China—it is similarly necessary to look at how the international and the domestic levels interact to shape Russia's nuclear politics and the extent to which the adoption of alternative security postures and arrangements by the other NWS might enable Russia to move towards a more liberal and democratic polity and achieve nuclear disarmament. The main body of this chapter investigates the various economic, political and social drivers of Russian nuclear possession in the post-Cold War era and outlines the changing role and importance of nuclear weapons in Russia's domestic and international politics. In addition, I discuss the obstacles to and opportunities for *institutional democratisation* pursuant to nuclear disarmament today, considering the current state of Russian public opinion and civil society engagement in nuclear matters in relation to Russia's authoritarian political model.

Given the history and scale of the Russian nuclear arsenal, this chapter is almost as long as the US chapter. I have endeavoured where possible to avoid repeating points concerning the two nation's Cold War rivalry and continuing adversarial relations. Important differences between the two case studies to note at the outset, which affect the size and character of the chapter, concern Russia's regime type and relative strategic power. Regarding the former point, several scholarly and other expert studies have portrayed Russia as moving further and further away from democracy towards illiberal authoritarianism over the course of the 2000s as a result of President Vladimir Putin's rule.

It would appear that the opportunities for *institutional democratisation* in this case are thus far more limited than in the US because Russia begins from a less conducive starting point. Yet there is also evidence showing that the Russian people support progressive action being taken on nuclear issues. These findings raise the question, which I shall examine below, of the extent to which the *guardianship* model of nuclear arms control and disarmament may be more appropriate and practical for Russia's political circumstances, at least in the short to medium term. In addition, there is the issue of Russian resistance to perceived Western meddling in its domestic affairs and those states in its near abroad. This raises another question, which I explore below: if nuclear disarmament requires *institutional democratisation*, how might the conditions arise internally, in the medium to long term, for Russia to move away from authoritarianism and towards democracy?

## 4.2 Mainstream and realist perspectives on the causes and consequences of Russian nuclear possession and disarmament

To begin with, it is necessary to again note that the mainstream and realist works reviewed in Chapter 2 mainly focus on US nuclear politics and strategy, for example, in terms of how an attack by the Soviet Union/Russia against the US or its allies could be credibly deterred. Far less space is therefore given over in these works to the Russian nuclear experience or the world as viewed from Moscow. Despite this, it is possible to apply the principles and ideas underlying the approaches reviewed in Chapter 2 to Russia, for example, by studying how Russian strategic analysts have applied a realist analysis in their work. In doing so we can observe what may be termed Russian perspectives on realist approaches to international relations, some of which mirror the descriptions and prescriptions of arguments promoting US interests.

Realist explanations of the origins of the Soviet nuclear programme focus on the threat posed by the US and its nuclear monopoly, which raised the possibility of a preventive nuclear first strike against the Soviet Union in the late 1940s and early 1950s.[1] Yet, as Geoffrey Roberts notes, whilst Stalin saw the bomb as a 'very important addition to his military arsenal ... it did not define the Soviet Union's postwar defence posture', because this was based on absorbing a NATO attack and then launching a 'land invasion of Western Europe'.[2] In addition to defensive concerns, the importance to the Soviets of attaining atomic power was stimulated by Stalin's desire for military strength to influence events on the world stage.

Stalin's decision to acquire nuclear weapons, two weeks after the US's atomic bombing of Japan in 1945, was taken, David Holloway argues, because he saw the 'use of the bomb as an anti-Soviet move designed to deprive the Soviet Union of strategic gains in the Far East and more generally to give the United States the upper hand in defining the postwar settlement'.[3] Connected to this, Craig and Radchenko argue, was Stalin's general mistrust of the West. The fact that Russia was 'beseiged by a capitalist world bent on destroying it' after WW2 was viewed by the Kremlin as the continuance of 'a long-standing tradition'.[4] These factors drove Moscow to devote immense resources to a crash programme that allowed a bomb to be tested in 1949—several years earlier than predicted by US intelligence.[5]

As noted above, by ending the US nuclear monopoly the Soviet nuclear programme was intended to strengthen the nation's bargaining position and great-power claims. Yet such a development was framed by the US State Department's influential NSC-68 report of 1950 as an essential part of the USSR's 'hostile designs' to spread Communism around the world. Moscow's 'formidable power' thus presented 'the gravest threat to the security of the United States'.[6] Such assessments of Soviet aims inform traditionalist Western perspectives on the Cold War, which emphasise the USSR's expansionist ambitions and military strength. In such accounts, generally speaking, domestic political factors are not ignored so much as they tend to play a lesser role.

After Stalin's death, Soviet strategy followed US policy in key areas, with conventional forces cut back and, Holloway argues, an 'increasing emphasis on nuclear weapons and on ballistic missiles as the means to deliver them'.[7] Thereafter, as retired Russian General Vladimir Belous explains, there emerged an approximate balance between both sides in strategic offensive weapons that contributed naturally to strategic stability and the concept of nuclear deterrence based on the central model of mutual assured destruction (MAD) that has never lost its topicality.[8]

Today, this 'topicality' continues based on the fact that Russia feels threatened by the US's far superior conventional military capabilities, which under the auspices of NATO, now reach close to its territory. The US also possesses an immensely powerful nuclear arsenal, which has been designed and developed to ensure the success of a pre-emptive attack on enemy forces. Given this aggressive military posture, Russian planners have to respond to the potential for a successful US nuclear first strike, which naturally prevents a more harmonious political relationship between the two nations.[9] Regarding nuclear disarmament, Russian experts argue that significant changes to the international political scene are essential if progress in this area between Moscow and Washington is to be possible. Notably, a 2010 article entitled 'Start a New Disarmament Plan' by the Russian 'gang of four' Yevgeny Primakov, Mikhail Moiseyev, Igor Ivanov and Evgeny Velikhov, argued that in the long term, a 'world without nuclear weapons is not our existing world minus nuclear weapons' but that this endeavour necessitated 'a thorough overhaul of the entire international system' that included the construction of a 'reliable mechanism for peaceful settlement of major and local international and border conflicts'.[10]

Russia's determination to retain its great-power status—seen by the Kremlin as having been shamefully lost in the 1990s—is another significant problem for disarmament efforts today. Evidence of these ambitions can be found in the 2013 *Foreign Policy Concept of the Russian Federation*, which outlines how the nation has 'a special responsibility for maintaining security in the world both on the global and regional levels'. According to this document, Russia pursues its goals via a 'multi-vector policy', which includes giving a central role to the United Nations and the development of a multipolar world order based on 'international law and principles of equality, mutual respect and non-interference in internal affairs of states'.[11]

Whilst a level playing field is called for at the international level, so that Russia sits as an equal at the top table of diplomacy, democratic principles are not applied at home. Given the authoritarian nature of the ruling elite, the domestic causes and consequences of nuclear disarmament are thus not considered in official or expert analysis, so that the only scenario entertained is one where Moscow ascends once again to international parity with the US. In the background of the Kremlin's strategic calculus also lies nuclear-armed China, with which Russia shares an extensive border.

Russia's global strategy is informed by a rejection of the 'new thinking' developed by former premier Mikhail Gorbachev and his circle at the end of the Cold War. Mikhail Tyspkin links the rejection of Gorbachev by contemporary Russian elites to their 'pursuit of great power status'. This urge, he claims, did not start with Putin, but materialised soon after the USSR's fall in 1991, when

> a consensus emerged across the range of Russian elites that (1) Gorbachev's 'new thinking', which discounted the use of different types of power in international relations, had failed, and (2) Russia's natural role in the international system is that of a major power.[12]

The memory of both the failure and betrayal of the Gorbachev years, where significant strategic concessions were made to the West without adequate reciprocation, and the Yeltsin presidency, where Russia is remembered as being disgracefully weak, are of particular importance in understanding the nationalism alive in Russia today and the new 'realism' in its international policy.

As things currently stand, the Kremlin is therefore strongly resistant to nuclear disarmament, so that whilst it publicly endorses the goal of nuclear zero, this is regarded as such a long-term aspiration as to be impractical. As Nikolai Sokov explains, when President Putin signed a law in 2010 on funding for upgrading the nuclear weapons complex he mentioned that the country would need its arsenal for the next '30-40-50 years'.[13] Moreover, democratisation—those from the Russian establishment might well argue—whether in pursuit of disarmament or other initiatives, is misguided given the evidence suggesting that the Russian public strongly supports the nation's nuclear status and foreign policy under Putin more widely. For example, a 2006 poll indicated that 76% of respondents believed that Russia 'needs nuclear weapons', with over 50% considering nuclear weapons to be the main guarantee of security.[14]

Sceptics may therefore, reasonably, use such data to support their claim that the domestic political barriers to nuclear disarmament in Russia are too high for this project to succeed in the near future. Given that nuclear weapons are one of the main currencies through which Moscow seeks to maintain its international standing, the process of devaluing these weapons is thus particularly challenging. This is also because proposals put forward by Russian experts almost exclusively place arms control and disarmament measures within a bilateral or multilateral framework involving an incremental step-by-step approach (the 'P5' process), managed by the NWS.

## 4.3 Critical perspectives on the causes and consequences of Russian nuclear possession and disarmament

In terms of the applicability of critical approaches—including the domestic politics model and the concept of *institutional democratisation*—to the Russian case study, it is useful to begin with the subject of regime type. Following the 1917 revolution Russia transitioned from an absolutist and autocratic monarchy to a self-proclaimed socialist republic. However, soon after any democratic element of the revolution was purged as Lenin and then Stalin implemented a brutal dictatorship.[15] Indeed, Dahl argues that Lenin's leadership of the Bolshevik party created a type of *guardianship* that had a 'unique ... claim to rule', namely, in the interests of the proletariat.[16]

In 1945 Stalin put a close associate of his, Lavrenti Beria, in charge of the Soviet nuclear programme. Despite the devastation of wartime, Beria created, according to Alexander Vershinin, 'a super-ministry with enormous resources and emergency powers' which eventually became 'a new industrial sector ... in the space of a few years—the atomic industry'.[17] Holloway notes that, as time passed, the atomic and defence industries and military in the Soviet Union 'became increasingly influential in the formulation of policy', in a way similar to the process that took place in the US.[18]

For Sergey Koulik, policies 'formulated under conditions of wartime emergency' gradually turned into 'a routine process of catching-up', given the US's military strength.[19] The USSR's 'military-industrial machine' was thus imbued with 'a strong momentum which was geared towards the pursuit of military competition for years to come'.[20] Whilst there was some truth in the official propaganda line that the Soviet bomb was a defensive measure to ward off imperialist aggression from the West, this narrative also hid the domestic reality of Soviet society, which was, for Koulik, stuck in the 'inertia of the military machine'—a fact that revisionist perspectives on the Cold War should recognise.[21] Moscow and Washington's Cold War rhetoric therefore presented a mirror image of the other side—with both claiming that their bomb was peaceful whilst the opponent's symbolised only nefarious intent. Such propagandising was routinely deployed at home and internationally. For the Kremlin, the bomb was a scientific and technological triumph that boosted its prestige, yet the Soviet leadership also felt the need to present its new weapon on the international stage as a symbol of peace and emphasise a preference for disarmament.[22]

Whilst the USSR tried to cultivate a positive image in the eyes of world public opinion and respond to criticisms of its foreign policy, domestically there was no need for such efforts. Thus, as April Carter explains, the absence of 'any direct popular or legislative pressure' on the Kremlin during the Cold War on arms control policies meant the Soviet leadership was never held accountable on this issue and also did not feel pressure to rein in its military programmes.[23] Yet this situation changed following Gorbachev's rise to power in 1985. For example, Bruce Russett makes the point that the 'internal character of the Soviet system moved importantly towards democracy during the 1986–1989 period'.[24]

According to Arbatov, this development ushered in a 'golden age' of 'civilian control and democratic accountability in their peculiar Soviet reforms', whereby academics and experts were involved in policy making, including 'the major disarmament endeavours of the time'. Arbatov credits this movement—with Gorbachev's support—for taming the military establishment, enabling the agreement of the 1987 INF Treaty, the 1990 Conventional Armed Forces in Europe (CFE) Treaty and START I.[25] Moreover, Cortright claims that Gorbachev and the European peace movements should be credited for 'demanding political change', which led to the Soviets offering 'sweeping nuclear reductions', including the elimination of all nuclear weapons at the 1986 Reykjavik Summit.[26]

Matthew Evangelista embellishes on these points in highlighting how the real significance of Gorbachev's administration was the realisation that Soviet actions, which were intended as defensive, at times appeared threatening to the West—even to those who 'were sensitive to Soviet security interests' such as 'members of European peace movements'.[27] This influenced the shift in Soviet foreign policy towards 'common security' as outlined in the Palme Commission report, which advocated international cooperation to achieve complete and general disarmament and the peaceful resolution of conflicts.[28]

Andrei Grachev, meanwhile, emphasises how Gorbachev was also one of a new generation of political leaders who 'were not prepared to accept uncritically the unlimited demands of the military-industrial complex, especially at a time when it had begun to devour vital parts of the body of Soviet society'.[29] The Soviet military-industrial complex (MIC) had, by the late 1970s, 'gained a position of almost unrestrained domination of the political and economic life of the country', making it immune from political control.[30] Importantly, particularly given our previous discussion of the influence of the MIC/Pentagon System in US nuclear weapons decision-making, Grachev also concludes that 'the two superpowers' engaged in the arms race as they were 'obviously motivated more by a common interest in preserving the status quo (including the positions of their respective military lobbies and industrial complexes) than by real defence concerns'.[31]

The preceding analysis suggests that the principal significance of Gorbachev for this study was his attempt to exert political control over nuclear weapons decision-making, hitherto dominated by military-industrial interests. Moreover, his brand of 'new thinking' prioritised disarmament as a means to reorient the Soviet economy away from military spending in order to provide more and better goods and services for its citizens and to reduce tensions with the West on the path to international cooperation between nations.

Cortright argues that Gorbachev's strategic concessions and sweeping economic and political reforms in tandem with the popular movements in Eastern and Western Europe presented a golden opportunity to realise nuclear disarmament and build 'more reliable structures of peace and international cooperation'.[32] Instead, the theory of 'cold war triumphalism', whereby military power broke the Soviet economy, embedded itself in US strategic thinking.[33] This led to a series of actions by the US that, Russian analysts argue, pushed their country into a corner and increasingly led to a reliance on the military and nuclear weapons for security as the 1990s wore on. For example, the authors of the Russian Institute of World Economy and International Relations (IMEMO) study *Russia and the Dilemmas of Nuclear*

*Disarmament* state that the continued existence of NATO as a military alliance following the dissolution of the Warsaw Pact, the incorporation of former pact members into NATO and the latter's subsequent expansion—breaking 'commitments previously undertaken at the highest level'—meant Russia was treated like a 'loser' in the Cold War, with its weakness taken advantage of for Western gain.[34]

Thus whilst there was an opening for strategic cooperation and the development of a security community between the West and Russia following the end of the Cold War, symbolised by Gorbachev's proposal for a 'common European home', this withered away as the US preferred enlargement to restraint.[35] Such expansionist policies during the 1990s and 2000s dealt a severe blow to the possibility of cordial US–Russian relations. Furthermore, as I explore in more detail below, US behaviour at this time contributed significantly to an increasing Russian reliance on nuclear weapons in defence and foreign policy, which, as several Russian analysts note, was not inevitable.

US–Russia relations are also important to study because of the influence Washington's behaviour has on the ability of different sub-national actors and groups in Russia to realise their policy preferences. As Gvosdev and Marsh note, there are different 'sectors and vectors' within Russia that create various competing foreign policies.[36] In the years following the end of the Cold War, US behaviour empowered more conservative actors, stimulating the revival of Russian nationalism. This is significant for the claim that *institutional democratisation* is necessary for nuclear disarmament because states with particular influence over Russian actions, such as the US, need to consider how their behaviour impacts on Russia's domestic politics—including supporting or undermining actors promoting military restraint and liberal, democratic reforms.

To argue that the West should be more sensitive in its dealings with Russia if it wants a more liberal and cooperative regime in Moscow is not to absolve the latter for its part in obstructing democratisation and disarmament nor to place all the responsibility for the status quo at the West's door. As Lilia Shevtsova argues, the main aim of the Russian political system at home is for ruling elites to maintain their monopoly of power.[37] Nuclear weapons are thus a part of the regime's survival strategy, so that, as Dmitri Trenin argues, since the end of the Cold War, in Russian domestic politics, 'control over nuclear weapons has become the ultimate symbol of presidential authority, an equivalent of the old-time scepter and the orb'.[38]

Rather than moving towards becoming a liberal democracy, the Kremlin has therefore centralised power and implemented a series of policies which led the Economist Intelligence Unit, from 2006 to 2015, to classify Russia as an 'authoritarian' state. Elsewhere, a 2017 Freedom House report gave Russia a score of 20/100 and classified it as in the bottom of seven ranks for political rights and in the second to last rank for civil liberties.[39] Notwithstanding the severe democratic deficit in Russia, examining the preferences of the Russian

public concerning security issues remains vital if we are to familiarise ourselves with the potential for progressive change.

As noted in the previous section, in addition to the consensus on foreign
and security policy amongst decision-makers, some evidence suggests that
the Russian public strongly supports the nation's nuclear status. Pavel Podvig
argues that the Russian public's support for its government's nuclear weapons
policies is largely

> a result of the lack of an open and informed discussion of national
> security priorities and policies that would involve independent voices.
> While there are non-governmental research organizations that are
> involved in the discussion of defence policies, there are no independent
> public organizations that would have nuclear weapons related issues on
> the agenda.[40]

Furthermore, according to one Russian military expert I interviewed, 'alternative voices are present but negligible, nearly anonymous'.[41] Despite the difficult political environment, there thus remain some Russian NGOs—with
connections to Western partners—that continue to work on these issues.

Whilst Podvig makes a strong case regarding why many in Russia back the
government's nuclear policy, apparently contradictory polling data suggests
that there is also significant support amongst Russians for arms control
and disarmament. For example, a 2005 poll by VTsIOM showed that 39%
of Russians supported nuclear reductions, but not to zero.[42] Similarly, a poll
two years later reported strong support amongst the Russian public for deep
cuts to the nation's arsenal—with 53% even favouring reductions below 400
nuclear weapons.[43] However, a survey in 2010 by VTsIOM struck a different
note, as its polling showed that 60% of Russians opposed further reductions,
with many believing New START had benefited the US more than Russia.[44]
Clearly these findings present a mixed picture of Russian attitudes to nuclear
issues, suggesting a preference for reductions alongside a continuing interest
in Russia remaining a nuclear possessor—a position influenced by a wariness
of Western intentions and a desire for the nation to retain a semblance of
great-power status.

In terms of the medium- to long-term changes required for Russia democratisation, Andreas Umland thus makes the point that

> Russia will become a law-ruled democracy when it stops seeing herself
> as a civilizational center engaged in a geopolitical struggle beyond her
> borders. Once the Russians discard the mirages of 'The Third Rome' and
> imperial greatness, they will finally become free.[45]

The argument that Russia's overblown ambitions need to be discarded if progressive and liberal reforms are to be realised raises several questions, which
I discuss further below. These include whether the Russian people or Russian

elites—or both—need to embrace this shift in mindset, what a new Russian identity might look like, how such a shift may occur (for example, through civil society movements), and how such a transition connects to disarmament action.

## 4.4 Russian nuclear politics in the post-Cold War world

In order to provide an overview of the different factors pushing and pulling Russia towards and away from reliance on nuclear weapons following the end of the Cold War, I will begin by outlining the former before turning to those developments potentially conducive to Russia reducing the salience of nuclear weapons in its security policy. The purpose of this investigation is to assess the argument that the underlying political relationship between Russia and the US has been the principal determinant of Russian dependence on nuclear weapons over the last two decades. In the following section I then consider the importance of other domestic, regional and international factors, and how they interact, as a means of assessing the explanatory value of *institutional democratisation* regarding Russian nuclear possession and disarmament.

As Dale Herspring notes, Boris Yeltsin's term as President (1991–1999) was experienced as a time of 'confusion and chaos' for those in charge of Russia's armed forces. With the economy in freefall, the resources available for conventional and nuclear weapons rapidly diminished, leading to Moscow's nuclear arsenal falling to a level 4 to 4.5 times below its Cold War peak during this decade.[46] Yet, as Sokov points out, whilst numbers were falling sharply, debate over the growing Russian reliance on nuclear weapons 'dominated the 1990s'.[47] For Sokov, this reliance was not fixed but fluctuated according to regional and international political events, 'peaking' in response to a combination of four variables: i) acute perception of external threat, ii) perceived absence of alternative means to ensure security, iii) perception of high utility of nuclear weapons, iv) cost-effective optimisation of military capability.[48] On the other hand, Russian scholars such as Arbatov and Vladimir Dworkin note that, 'despite serious differences on some issues' during the 1990s and early 2000s, there were also several constructive cooperative initiatives undertaken between Russia and the US.[49]

According to several sources, Yeltsin had made overtures to NATO regarding potential Russian membership 'as a long-term political aim' in December 1991, but received no response.[50] Instead, as Rumer posits, a weak Russia, transitioning from Soviet authoritarianism to democracy, was 'bypassed' by the West.[51] At this time, Western planners saw an opportunity to maximise their power, assuming that Russia would eventually acquiesce in the post-Cold War international order they were creating. Moreover, Rumer notes, moves to expand NATO and 'secure Russia's periphery' would also ensure that, if Russia re-emerged 'as a threat to Europe, as either a totalitarian or a failing state', it could be contained.[52] This path was chosen despite warnings in the late 1990s from a range of eminent US statespeople and political commentators that NATO expansion would lead to an adverse reaction in Russia and would constitute a 'policy error of historic proportions'.[53] Indeed, as Alexander Konovalov argued at the time, the result of these policies was that many Russians began to think that the US 'needs a weak, humbled Russia, perhaps as a source of cheap labor and raw materials'.[54]

For their part, Russian military strategists responded to the far superior NATO conventional military forces they saw as threatening Russian borders in 1996–1997 by emphasising the uses of tactical nuclear weapons (TNW), a development which 'propelled nuclear weapons into the center of attention and created a perception of their high utility' according to Sokov.[55] The next crisis between Russia and the West took place in 1999 with NATO's bombing of Serbia. Sokov describes how a meeting of the Russian Federation Security Council during the Kosovo war, headed by Putin, decided to 'enhance reliance on nuclear weapons in a departure from all documents adopted in the 1990s'.[56]

The primary innovation at this time was the new mission given to nuclear forces—deterrence of limited conventional wars.[57] The significance for Russia of the Kosovo war and the subsequent invasion of Iraq in 2003 was that it showed how the US could use force without the authorisation of the United Nation's Security Council. Moreover, Bill Clinton's decision in 1999 to go ahead with a national missile defence system, and amend the ABM Treaty accordingly, was viewed by Moscow as evidence that Washington was seeking the ability to launch a large-scale attack or even gain a first-strike capability.[58]

With Putin having assumed the Russian Presidency, the National Security Concept and the Military Doctrine of 2000 enshrined the new role of nuclear weapons outlined above. The reliance on nuclear weapons would be a temporary measure until conventional forces—particularly precision-guided munitions and missile defence—were developed.[59] In this sense nuclear weapons were a cost-effective means of deterring NATO's quantitatively and qualitatively stronger forces.[60] Yet the many problems with conventional modernisation—frequent delays, corruption by officials, inadequate production equipment and skilled personnel—were at least as great as those with nuclear weapons, meaning that progress was far more sluggish than planned despite the significant increase in spending.[61] The Kremlin is now also faced with the problem of a struggling economy (a situation made worse by Western sanctions following the annexation of Crimea) at a time when it plans further increases in military expenditure, leading to a greater reliance on defence supplies from abroad.[62]

Thus despite the attempts at military reform and the 2020 State Armament Programme, which aimed to equip the military with 30% of new arms by 2015 and 70% by 2020, Russia is unlikely to ever close the gap with China, the US and NATO, ensuring that its reliance on nuclear weapons will continue 'indefinitely'.[63] Several scholars agree that the outlook for Russia's conventional military is bleak—it now stands on the 'precipice of irrelevance' so that the nation may be classified as a 'strategic backwater'—according to a Washington-based analyst I spoke to.[64]

Whilst Beijing's small nuclear arsenal does not strike fear into Russian hearts, according to one Moscow-based expert I consulted, the superiority of its large and well-armed military cannot be ignored and remains in the background of Russia's nuclear planning.[65] Indeed McDermott identifies Moscow's conventional inferiority as a key driver of its continued possession of tactical nuclear weapons.[66] Similarly, Sokov has described how Russia partly retains its nuclear arsenal 'just-in-case' China 'becomes a foe ... or attempts to transform Russia into a subordinate power'.[67]

As noted above, whilst Russia reacted negatively to NATO expansion and overseas aggression, Arbatov and Dworkin identify several cooperative ventures that have been proposed or launched, to varying degrees of success, over the past two decades.[68] Moreover, two initiatives from 1997 indicated that a warmer relationship between Russia and the West was possible. The NATO–Russia Founding Act promised to ease concerns over NATO expansion and the US–Russia Helsinki Summit suggested that Russia's concerns over long-range conventional air-launched cruise missiles and sea-launched cruise missiles would be addressed in future arms control negotiations.[69] Whilst Washington's post-Cold War triumphalism and NATO's expansion were key factors behind the inability to reshape US–Russian relations, Sokov also suggests that Moscow could have been more 'constructive' and less 'passive' in its dealings with the West.[70] Today, as several commentators have argued, one area where all of the major powers could make a contribution is in making clear their intentions over future plans for military modernisation.[71]

With regards to Russia's relationship with Europe, one of the key strategic concerns is the 1990 CFE Treaty, which created the framework for conventional arms control and stability across the continent. Yet today this system is broken, with Russia having taken the drastic step of suspending its participation in 2007. In addition to requesting NATO members to ratify the Adapted CFE, Moscow made this move out of a concern that CFE 'flank limitations' could increase Russian vulnerability following Georgia and/or Ukraine joining NATO, a fear which intensified following the 2008 Georgia conflict.[72] In 2011 the US suspended its cooperation with Russia on CFE matters. Subsequently, amid tensions with the West over Ukraine, Moscow announced in March 2015 its final withdrawal from the CFE regime, NATO having not met its conditions for remaining a member.

Subsequently, as we have seen, Russia annexed Crimea, engaged in nuclear signalling—and intervened in Syria in its first out-of-area operation since the end of the Cold War. Yet these actions should be seen as acts of frustrated defiance and a demand for inclusion in settling global affairs—following

Moscow's patience with Europe and the US running out—rather than neo-imperialism. More recently, underlying tensions between Russia and the US have had significant impacts in the arms control arena. As discussed in Chapter 3, in February 2019 the US declared a suspension of its obligations under the INF Treaty, claiming that Russia's new cruise missile, the 9M728, was in violation of the rules. Russia denied this, pulling out in response and making the counter claim that the US deployment of a missile defence system capable of launching offensive missiles—the Mark 41 Vertical Launch System—is itself in material breach of the treaty.[73]

With the Kremlin already feeling more vulnerable to a decapitation strike by the US, or a disarming first strike, the end of the INF Treaty could see Moscow increase intermediate-range ballistic missile deployments and other new systems. Kristensen and his co-authors also predict that Russia will 'undertake countermeasures that would further increase the already danger-ously high readiness of Russian nuclear forces', raising the risk of nuclear weapons being used, including 'in response to early warning of an attack—even when an attack has not occurred'.[74] Notwithstanding such moves, as Squassoni notes, 'senior Russian officials' continue to 'express their willing-ness' to extend New START until 2026, despite the ongoing imposition of US sanctions following the Ukraine crisis.[75]

Russia's apparent commitment to New START exists alongside its nuclear modernisation plans, which analyst David Lonsdale has described as 'con-siderable', noting that it includes 'the RS 28 SARMAT super-heavy ICBM, PAK DA Stealth Bomber, Status-6 nuclear torpedo, Avangard hypersonic glide vehicle, and a nuclear-powered cruise missile'.[76] Whilst significant, such modernisation needs to be put into context in order to understand the stra-tegic implications. First, as discussed above, the Russian perspective is that it remains at a notable disadvantage in terms of the overall military balance with NATO, especially given the US's advances in BMD and conventional strategic weapons, which together pose a serious threat to Russia's nuclear arsenal, par-ticularly as it reduces in size over time.[77] This situation explains why President Putin stated in 2012 that his nation would develop 'high-precision weapons' in order to 'overcome any missile defence system and protect Russia's retali-ation potential'.[78]

Secondly, regarding Russia's development of new military technolo-gies, Rupal Mehta argues that Russia and China have 'upped the ante' by developing hypersonic capabilities able to carry nuclear warheads, whereas the US equivalent has been 'primarily conventional'.[79] However, Woolf notes that Moscow's Avangard system 'does not create a new threat to the United States or new issues for nuclear deterrence', so that discussions about Russian and Chinese hypersonic weapons should be interpreted 'as a competition in the development of new technologies' rather than a 'pending arms race'.[80] In addition, Michael Horowitz argues that, whilst Russia is 'investing heavily in military robotics and autonomous systems' to close the military gap with the US, it is unclear what 'actual progress' has been made.[81]

### 4.5 Domestic actors and interests driving Russian nuclear weapons decision-making

Having reviewed the significant ways in which the post-Cold War international security environment has shaped Russian nuclear weapons decision-making, we may now consider the domestic obstacles to and opportunities for nuclear disarmament, including how the domestic and international levels interact, and further assess the explanatory power of *institutional democratisation* for this case study. According to an experienced Russian analyst I interviewed, top Russian decision-makers value nuclear weapons most highly out of the five NWS because they have specific domestic political uses. In his opinion, nuclear weapons are a means by which the elite can 'manage the domestic population' as they are a potent symbol of the nation and the military's greatness, thus 'preserving the population's self-esteem'.[82] Similarly, Tsypkin has noted how both Yeltsin and Putin used nuclear missiles as 'political theatre' to maintain an image of strength amongst the Russia public.[83] These weapons thus provide 'stability', give the impression that the nation is still a leader in scientific and technological achievement and act as a distraction from the years of economic turmoil following the demise of the Soviet Union.

According to the Russian analyst just mentioned, there is thus 'no point' considering the question of nuclear disarmament under the current Russian leadership, as any serious moves in this direction would 'undermine its power base' and amount to 'political suicide'.[84] Karaganov has made similar arguments, stating that the idea of 'nuclear zero' is 'not only unrealistic, but outright dangerous', so that the only point of arms control for Russia is as a bargaining tool to build trust and transparency between the great powers.[85]

If this formulation is correct and the current ruling regime in Moscow has become inextricably linked with the possession of nuclear weapons, it is necessary to explore how the regime's power might be weakened so that alternative political groupings that are supportive of nuclear disarmament might emerge. One of the main problems we face here is that the political atmosphere in Russia is such that discussions of nuclear weapons by Russian analysts working in that country only rarely criticise domestic political decisions, presumably because to do so and actively promote nuclear disarmament would challenge the Kremlin's authority and thus carry personal and professional risks. It is worth emphasising, however, that Russia retains substantial expert knowledge, both governmental and non-governmental, concerning nuclear arms control and disarmament, stemming from the continual negotiations in this area since the 1960s.[86]

There are thus two alternatives available to decrease Russian reliance on nuclear weapons according to Sokov. The first, which has already been discussed, is for Russia to successfully build up its conventional forces so they can substitute for nuclear deterrence. The second is that Russia's security situation is

> improved just enough to facilitate a change in the domestic political lineup so that 'pro-nuclear' groups do not hold the veto over decision making on this issue. Then, if the political leadership decides to minimize reliance on nuclear weapons, it will be able to do so.[87]

This latter argument has much merit but does not address whether Russian elites should rein in their regional and international political ambitions or whether and how domestic political change could occur, leading to a more progressive government. As already asserted, it is therefore necessary to also consider the possibility of domestic political change occurring, with or without external developments conducive to disarmament.

More broadly, contemporary observers of Russia describe it as a 'managed' or 'vertical' democracy, with immense power concentrated in the President's office.[88] Putin has occupied this office for four non-consecutive terms since 2000, handling the powerful 'financial-industrial clans' upon which the Russian political system 'uneasily' rests, according to Tsypkin.[89] As analysts at IMEMO argue, after 2000, Moscow chose to 'insure its sovereignty and centralized rule by building an authoritarian political regime on the basis of carbon-export economy', a course 'which required a notion of immanent external threat as one of the instruments of consolidation'.[90] This centralised rule is particularly tight when it came to national security policy, including nuclear weapons, which several analysts note became far more restricted under Putin's rule decision-making than his predecessor as he 'built a decision-making pyramid that consolidated power in his hands'.[91]

For Arbatov, nuclear weapons are thus the section of the Russian state least influenced by 'civilian control and democratic accountability'.[92] This has led to a lack of understanding of the basics of nuclear weapons and arms control amongst much of the Russian political class.[93] Moreover, if military spending and procurement are not subject to more transparency, with some oversight by the Russian parliament (the Duma), media and general public, it is doubtful whether Russia will be able to achieve effective modernisation and reform in this area.[94] Yet at present, for Boris Kagarlitsky, the Russian state is only 'pseudo-parliamentary' because the Duma, when established, 'had no power' and a true multi-party system has not developed.[95]

Certain public opinion polls from recent years appear to show that Russians are generally supportive of their government's nuclear weapons policy. For example, one of the few in-depth polls on this subject from 2000 found that 40% of Russians believed that nuclear weapons gave their country 'political might', though another poll from 2005 suggested that Russians believed that several other factors, such as human rights and culture, were of greater consequence.[96] More generally, a 2007 poll showed that 63% of Russians support the aim of eliminating all nuclear weapons, with 66% wanting their government to 'do more to pursue this objective'.[97]

Importantly, there is a wider domestic function of nuclear weapons for the political elite that controls them so tightly, something often unmentioned and which does not appear in official documents—namely regime survival. As one Russian political analyst I interviewed suggests, nuclear weapons can deter Western attempts to 'get too involved in Russia's affairs'. Nuclear weapons therefore act as an 'ideological defence' for the Kremlin, which has recently seen regimes without nuclear weapons being toppled either by the West or with the latter's support in the Middle East and during the Arab Spring.[98] Nuclear weapons, in this sense, thus act as a tool to bind the elite to the people through a nationalist discourse whereby the 'managers of democracy' in the ruling United Russia party protect the Russian nation against external threats, which it has itself sometimes over-hyped for domestic political gain. This may in the short term provide benefits to the current occupants of the Kremlin but cuts against Russia's real long-term security interests and needs.

Thus, as Nikolai Zlobin avers, the Kremlin has cynically created 'a besieged-fortress mentality in the minds of the people' by rallying them against 'an outside enemy'.[99] As a result, global public opinion polls consistently show the Russian public to be among those with the most negative views of the US.[100] The result has been that Cold War stereotypes, which had never been wholly discarded, were revived as conflict outpaced cooperation and reactionary forces in the US and Russia found new soil in which to grow. It must be said, however, that until recently this relationship was more beneficial to Russian than American hawks given that, as we have seen, the latter principally turned to threats from terrorism, developing powers and a rising China to justify maintaining its huge military establishment after the fall of the Soviet Union.

Yet it remains the case that (albeit now to a different degree than during the Cold War) at times of tension the pro-militarist and pro-nuclear camps in both capitals feed off another in a kind of symbiotic relationship. This is because certain sections of the elite in both Russia and the US have a mutual bureaucratic and institutional interest in maintaining control and influence over national policy. Thus, according to Tsypkin and Anya Loukianova, the Russian defence industry supports 'continued, even enhanced reliance on nuclear weapons' and also largely opposes US–Russian cooperation because of 'both the traditional perceptions inherited from the Soviet period and the acute competition in global arms markets'.[101] The Russian military-industrial base therefore needs sections of the US's political establishment to engage in the game of demonisation, threats and sabre-rattling. This cycle of enmity benefits militarists and nationalists in both nations, so that US proponents of NATO members continuing to host TNW can justify this by pointing to an aggressive, rearming Russia as a dangerous belligerent, whilst US proposals for stationing BMD systems in Eastern Europe provoke 'outrage' in Moscow, fostering an 'atmosphere of grievance' which strengthens its ruling elite.[102]

The image of an all-conquering and belligerent Russia has therefore recently been conveyed to the world despite the fact that its economy is stagnating, its political model is unstable and its military is relatively weak.[103] With regards to the latter, Russia possesses only limited power projection capabilities—with a regional rather than global reach—and is surrounded by nations with more sophisticated, hi-tech weaponry.[104] Moreover, according to a 2013 report by the Natural Resources Defense Council 'the conventional military balance now favors NATO over Russia by a factor of 3 to 1', whilst 'in military expenditures, the ratio of NATO to Russia is 20 to 1'.[105]

One method not mentioned here, but by which Russia would be able to divert funds to much needed 'modernisation' and invest in social goods and services, is to cut spending on its nuclear weapons programmes and military forces more generally. Importantly, the burden of total military spending on Russia's economy is roughly twice that of other countries—excluding the US.[106] Reorienting the economy away from wasteful military expenditure should therefore be a priority, given that, according to the Organization for Economic Cooperation and Development (OECD), Russia has 'poverty and income inequalities well above the OECD average'.[107]

Separately, Tsypkin and Loukianova suggest that, if Russia's strong position as a 'leader' in nuclear energy and proponent of peaceful nuclear energy were 'gradually emphasized' above its status as a NWS, then this 'trade-off' might support initiatives to 'engage Russia on disarmament'.[108] Any shift to a new economic and political regime that is supportive of nuclear disarmament will significantly depend on Russia's external security environment and how acceptable such changes are to Russian society at large. One key question here, as Rumer highlights, is how amenable Russians might be to further social and political 'upheavals'. Given their recent history, this author argues, Russians currently prefer stability to the 'uncertainties' of liberal democracy, despite the significant grievances held against their own government.[109]

Elsewhere, prominent Russian analysts emphasise the importance of the country engaging in liberal and democratic reforms, but have also urged caution regarding the prospects for such a transformation. For example, Kagarlitsky has argued that the Russian people 'are not ready for democracy' but 'do not want dictatorship', a situation which arose with the 'decay of society' after the fall of the USSR and which is 'aggravated in turn by the bankruptcy of politics'.[110] He therefore proposes that the 'historic task' for Russia, if the nation's economic and political life is to survive, will be to 'search for new forms of social being', which will emerge from and support a self-organising and politically engaged citizenry.[111] Signs of such a democratic movement could be seen the anti-corruption protests of 2011, so that, as Perry Anderson observes, 'widespread opposition' to Putin exists 'in the centre of the country'.[112] Yet, for former presidential aide Gleb Pavlovsky, a credible rival to Putin, who will be able to maintain existing living standards, has yet to emerge.[113] Furthermore, in March 2020 Putin supported a measure that could allow him to remain Russian president until 2036, with the Kremlin pointing to the 'extreme turbulence' in the world as justification for the far-reaching constitutional changes.[114]

Looking forward, Rumer therefore argues that Americans and Europeans need to realise that the emergence and development of democratic and progressive forces in Russia require them to 'engage, explain, listen and watch carefully, and help wherever help is sought and possible to give'.[115] To this extent, the path to cooperation on nuclear disarmament will depend on wider forms of cooperation and confidence-building requiring patient engagement between nations over the long term. Furthermore, this type of initiative will likely only be possible if the costs of inaction and the benefits of success are clearly explained to the American and Russian publics by political leaders, the media and civil society, in order to harness the energy required to break through the current impasse and sustain such efforts over time.

## 4.6 Conclusion

This chapter has explored a range of views on Russian nuclear politics in order to ascertain the explanatory power of mainstream, realist and critical approaches to the causes and consequences of nuclear possession and disarmament. For the Russian case study, the security model is most persuasive when outlining the high barriers to nuclear disarmament driven by Russia's perception of the US and NATO as an existential threat. Nuclear weapons also continue to be highly valued because the Kremlin and, to an extent, the general populace, still see their nation as a great power and regional hegemon seeking independence from Western containment. Moscow's nuclear deterrence strategy is therefore a response to a national sense of Western encirclement as well as the need to ensure regime survival. In addition, as previously discussed, nuclear possession plays an integral role in Russia's domestic political system as a symbol of both the ruling elite's domination over society and continuity with the Soviet era, when the nation was a world superpower— which is in line with the expectations of *institutional democratisation*.

Given that Russia is an authoritarian state dominated by one party—a *guardianship* in and of itself—it is clear that there are thus greater barriers to *institutional democratisation* than for the nuclear possessors that are formally liberal democracies. As such, despite the current regime showing signs of vulnerability given recent crises and troubles at home and abroad, it is likely that Russian democracy and moves towards nuclear disarmament will take some time to emerge. In the short term it will therefore be vital for the guardians of Russia's nuclear arsenal to do their part in maintaining arms control agreements and ensuring strategic stability with the US, to prevent conflict and thus avert a potential nuclear war.

Despite the fractured nature of relations between Moscow and Washington today, the period from the fall of the Soviet Union to the present can be seen as one of a series of lost opportunities for the causes of both *institutional democratisation* and nuclear disarmament. Although Reagan and Gorbachev came close to agreeing to the abolition of nuclear weapons at Reykjavik, the latter's new thinking, driven by a spirit of cooperation and nonviolence on the international front and liberalising moves at home, opened up the potential to permanently demilitarise relations between East and West. Some progress was made around this time, with important arms control agreements offering hope that a new and lasting political and security settlement could be reached with the end of the Cold War. Yet, as discussed in Chapter 3, the ABMT and INF treaties have been jettisoned and the future of the New START and the CFE treaties is uncertain—just when such initiatives are most needed.

The current malaise can be explained by the fact that, following the demise of the Soviet Union, a triumphalist US and NATO—throughout the 1990s and into the 21st century—chose to largely exclude Russia whilst building the new international order, instead acting in ways which Russians found particularly threatening. This led a weakened and humiliated Russia, now led by Vladimir Putin, to cling to its nuclear weapons as one of the few remaining sources of strength in a unipolar world ruled by force. Gorbachev's idealism was thus well and truly dead, having been replaced by a realism of short-term tactics aimed at protecting Russian sovereignty from external meddling. The power of the President and his circle of oligarchs was thus consolidated whilst the prospects for social democracy withered away.

This situation clearly benefited nationalist and militarist forces in Russia who were able to rally popular support for increased spending on conventional and nuclear weapons as part of a renewed emphasis on finding military solutions to political problems, all of which boosted their own institutional power bases. Whilst, rhetorically, US and Russian political leaders professed to be developing a new political partnership, in reality Moscow increasingly saw NATO's actions as a direct danger and acted accordingly. Behind the scenes, US and Russian officials did continue meeting to discuss issues of strategic stability and manage their nuclear standoff, with bureaucrats exchanging detailed information about several aspects of their arsenals during arms control negotiations. Yet the responsibility each nation has to achieve nuclear disarmament required them to commit to a long-term political partnership to deal with the mutual threat of nuclear war, something that neither of these nation's leaderships—with their particular brands of elite economic domination, and prominent nationalist and militarist cultures—are or were capable of.

Russian leaders thus prefer to retain a sizeable nuclear arsenal as this enables them to sit alongside their US counterparts as equals on the world stage. The Kremlin also faces virtually no domestic pressure within Russia that may cause it to reconsider the costs and benefits of nuclear weapons and the policies that govern them. Whilst there is evidence that significant sections of the Russian public would support moves to further reduce their nuclear arsenal, there is no domestic institutional focus for their preferences, either in the media, civil society or parliament. The current dire state of relations with the US also means that disarmament talk struggles to find prominent supporters in Russia, with unilateral measures completely off the table. In any case, information on the Russian nuclear weapons system is scarce and rarely discussed publicly—except to drum up patriotic fervour.

Yet were debate concerning the social, environmental and financial costs and risks of these weapons to enter public discourse it is possible that opposition to them could, as seen in Western democracies, rise accordingly, opening up a new political space for dissent and alternative security policies to be proposed. Given the immense costs and risks to Russia of modernising its nuclear arsenal, the current moment should present an opportunity to raise fundamental questions about the necessity of Russia retaining its nuclear weapons, so that it moves towards zero rather than a more resilient deterrent—militarily and politically—at lower numbers. Moreover, the weighty technical and diplomatic knowledge concerning arms control and disarmament resident in Russian governmental and non-governmental bodies could be harnessed to support such efforts as part of moves in the direction of *institutional democratisation*.

However, the current worldview of Russia's decision-making elites, including the perception that calls for Russian nuclear disarmament are a trick or plot by the West, means that it is likely that the elimination of the Russian nuclear arsenal will only take place when a new, far more democratic and liberal-minded regime is in place. How this might happen is clearly a complex question for the medium to long term, yet it is possible to identify some measures that can be taken in the short term to cultivate the development of such progressive forces. For example, given the internal contradictions and many weaknesses within its economic and political model, Russia—meaning the nation and its citizens rather than the elite alone—needs to find a new basis for security and an alternative purpose beyond achieving 'great-power status', a goal that primarily benefits those in and connected to the Kremlin. To keep channels open for dialogue and diplomacy, which may include discussions for how a more settled relationship between East and West can be achieved, major European powers will therefore need to maintain cordial relations with Moscow.

Yet the sanctions imposed on Russia following the Ukraine crisis have damaged an already failing economy, leading nationalists to call for Russia to distance itself further from the West. This may appear attractive to the purveyors of 'vertical democracy' as a short-term means to harness nationalist sentiment, but will, in the long run, only further undercut the possibilities for the modernisation of Russia's infrastructure and the health and wellbeing of its populace. Moreover, if Russians perceive their lot to be worsening over the next decade, it will be harder for the ruling elite to maintain their hold on power if discontent returns to the streets of Russian cities, which could result in a more unpalatable regime—with an even greater commitment to military might and nuclear weapons—attaining power.

Looking forward then, if Russia is to move away from reliance on nuclear weapons it will need to be included in cooperative proposals coming from the

US and Europe that recognise Russia's legitimate security fears and concerns. As discussed in the previous chapter, recent developments in US–Russia arms control have not, to say the least, inspired optimism for such a shift of priorities. Yet it is worth noting that other middle powers, such as Germany, might play a constructive role in mending the relationship between Russia and the West. For example, Germany and other influential states on good terms with Russia could put political capital into revisiting some of the economic and security proposals for Russo-European cooperation made in recent times. Looking more widely, the promise of a truly multipolar world order where Russia feels itself to be an equal and is integrated economically and politically with both the rising powers of Brazil, India, China and South Africa, but also the other major Western nations, could open up new possibilities for international cooperation and the construction of security communities that delegitimise the threat and use of force as instruments of state policy.

## Notes

1  Pincus, Walter (1983) No First Strike, www.washingtonpost.com, 16 May.
2  Roberts, Geoffrey (2006) *Stalin's Wars: From World War to Cold War, 1939–1953* (New Haven, CT: Yale University Press), 364.
3  Holloway, David (2012) Nuclear Weapons and the Escalation of the Cold War, 1945–1962, in Leffler, Melvyn P. and Westad, Odd Arne, eds., *The Cambridge History of the Cold War*, vol. 1, *Origins* (Cambridge, Cambridge University Press), 377.
4  Craig, Campbell and Radchenko, Sergey (2008) *The Atomic Bomb and the Origins of the Cold War* (New Haven, CT: Yale University Press), 165.
5  Ibid., 378.
6  US State Department (1950) *NSC-68: United States Objectives and Programs for National Security*, www.mtholyoke.edu, 14 April.
7  Holloway, Nuclear Weapons and Escalation, 386.
8  McDermott, Roger N. (2011) Russia's Conventional Armed Forces: Reform and Nuclear Posture to 2020, in Blank, Stephen J., ed., *Russian Nuclear Weapons, Past, Present and Future* (Carlisle: Strategic Studies Institute), 138.
9  Kristensen, Hans M., Norris, Robert S. and Oelrich, Ivan (2009) *From Counterforce to Minimal Deterrence: A New Nuclear Policy on the Path toward Eliminating Nuclear Weapons* (Washington, DC: Federation of American Scientists), 26–27.
10  Primakov, Yevgeny, Moiseyev, Mikhail, Ivanov, Igor and Velikhov, Evgeny (2010) Start a New Disarmament Plan, www.in.rbth.com, 22 October.
11  Russian Federation (2013) *Concept of the Foreign Policy of the Russian Federation*, www.mid.ru.
12  Tsypkin, Mikhail (2009) Russian Politics, Policy-Making and American Missile Defence, *International Affairs*, 85(4), 787.
13  Sokov, Nikolai (2011) Nuclear Weapons in Russian National Security Strategy, in Blank, *Russian Nuclear Weapons*, 188.
14  Akhtamzyan, Ildar (2006) *Attitudes in the Russian Federation towards WMD Proliferation and Terrorism* (Moscow: Human Rights Publishers), 16.
15  Fitzpatrick, Sheila (2008) *The Russian Revolution* (Oxford: Oxford University Press), 41.

16  Dahl, Robert A. (1989) *Democracy and its Critics* (New Haven, CT: Yale University Press), 53.

17  Vershinin, Alexander (2017) Why Did the Soviet Union Develop its own Atomic Bomb? www.rbth.com, 24 March.

18  Holloway, Nuclear Weapons and Escalation, 396.

19  Koulik, Sergey (1991) The Soviet Union: Domestic and Strategic Aspects of Nuclear Weapon Policy, in Carp, Regina Cowen, ed., *Security with Nuclear Weapons? Different Perspectives on National Security* (Oxford: Oxford University Press), 125.

20  Ibid., 139.

21  Ibid., 140.

22  Barghoorn, Frederick C. (1964) *Soviet Foreign Propaganda* (Princeton, NJ: Princeton University Press), 111.

23  Carter, April (1989) *Success and Failure in Arms Control Negotiations* (Oxford: Oxford University Press), 26.

24  Russett, Bruce (2011) *Hegemony and Democracy* (Abingdon: Routledge), 99.

25  Arbatov, Alexei (2010) Russia, in Born, Hans, Gill, Bates and Hanggi, Heiner, eds., *Governing the Bomb: Civilian Control and Democratic Accountability of Nuclear Weapons* (Oxford: Oxford University Press), 54–55.

26  Cortright, David (2008) *Peace: A History of Movements and Ideas* (Cambridge: Cambridge University Press), 150.

27  Evangelista, Matthew (1999) *Unarmed Forces: The Transnational Movement to End the Cold War* (Ithaca, NY: Cornell University Press), 306.

28  Palme, Olof (1982) *Common Security: A Programme for Disarmament* (London: Pan Books).

29  Grachev, Andrei (2008) *Gorbachev's Gamble: Soviet Foreign Policy and the End of the Cold War* (Cambridge: Polity Press), 38.

30  Ibid., 16.

31  Ibid., 19.

32  Cortright, *Peace*, 150, 323.

33  Ibid., 149.

34  Arbatov, Alexei, Dvorkin, Vladimir and Oznobishchev, Sergey, eds. (2012) *Russia and the Dilemmas of Nuclear Disarmament* (Moscow: IMEMO RAN), 97–102.

35  Gorbachev, Mikhail (1989) Europe as a Common Home, address given to the Council of Europe, Strasbourg, 6 July.

36  Gvosdev, Nikolas K. and Marsh, Christopher (2014) *Russian Foreign Policy: Interests, Vectors and Sectors* (Los Angeles, CA: Sage), xiii.

37  Shevtsova, Lilia (2012) Russia's Choice: Change or Degradation, in Blank, Stephen J., ed., *Can Russia Reform? Economic, Military and Political Perspectives* (Carlisle: Strategic Studies Institute), 2.

38  Trenin, Dmitri (2005) *Russia's Nuclear Policy in the 21st Century Environment*, www.ifri.org, Autumn, 7.

39  Economist Intelligence Unit (2016) *Democracy Index 2016: Revenge of the 'Deplorables'*, ww.eiu.com, 39; Freedom House (2017) *Populists and Autocrats: The Dual Threat to Global Democracy*, https://freedomhouse.org, 23.

40  Podvig, Pavel (2012) Russian Federation, in Acheson, *Assuring Destruction Forever*, 7.

41  Interviewee: D.

42  Podvig, Pavel (2005) Russian Public Opinion on Nuclear Weapons, www.russianforces.org, 5 August.

43  World Public Opinion (2007) American and Russian Publics Strongly Support Steps to Reduce and Eliminate Nuclear Weapons, www.worldpublicopinion.org, 9 November.

44  RIA Novosti (2010) Most of Russians Against Nuclear Disarmament, www.en.ria.ru, 15 July.

45  Umland, Andreas (2012) Will Russia Become a Democracy? http://nottspolitics.org/, 4 March.

46  Herspring, Dale R. (2011) Russian Nuclear and Conventional Weapons: The Broken Relationship, in Blank, *Russian Nuclear Weapons*, 3, 8.

47  Sokov, Nikolai (2002) Why Do States Rely on Nuclear Weapons? The Case of Russia and Beyond, *The Nonproliferation Review*, Summer, 101.

48  Ibid., 105–106.

49  Arbatov, Alexei and Dworkin, Vladimir (2005) *Revising Nuclear Deterrence* (College Park, MD: University of Maryland), 8.

50  Friedman, Thomas (1991) Soviet Disarray; Yeltsin Says Russia Seeks to Join NATO, www.nytimes.com, 21 December; Straus, Ira L. (1997) *The Evolution of the Discussion on NATO-Russia Relations*, www.fas.org, February; Trenin, Dmitri (2007) *Getting Russia Right* (Washington, DC: Carnegie Endowment for International Peace), 71.

51  Rumer, Eugene B. (2007) *Russian Foreign Policy beyond Putin* (London: Routledge), 18.

52  Ibid., 19.

53  Friedman, Soviet Disarray; Arms Control Association (1997) Opposition to NATO Expansion, www.armscontrol.org, June–July.

54  Schell, Jonathan (1998) *The Gift of Time* (London: Granta), 158.

55  Sokov, Why Do States Rely, 103–104.

56  Sokov, Nuclear Weapons in Russian Strategy, 205.

57  Sokov, Nikolai (2004) *Russia's Nuclear Doctrine*, www.nti.org, 1 August.

58  Sokov, Why Do States Rely, 102–104.

59  Sokov, Nuclear Weapons in Russian Strategy, 249.

60  Sokov, Why Do States Rely, 105.

61  Westerlund, Frank (2012) The Defence Industry, in Pallin, Carolina Vendil, ed., *Russian Military Capability in a Ten Year Perspective* (Stockholm: FOI, Swedish Defence Research Agency), 76; Herspring, Dale R. (2011) Russian Nuclear and Conventional Weapons: The Broken Relationship, in Blank, *Russian Nuclear Weapons*, 26.

62  Myers, Steven Lee, Becker, Jo and Yardley, Jim (2014) In Putin's System, Closest Friends Reap the Richest Rewards, *New York Times International Weekly*, 5 October.

63  Oxenstierna, Susanne and Bergstrand, Bengt-Goran (2012) Defence Economics, in Pallin, *Russian Military Capability*, 58; Sokov, Nuclear Weapons in Russian Strategy, 249.

64  Interviewee E; Goure, Daniel (2011) Caught between Scylla and Charybdis: The Relationship between Conventional and Nuclear Capabilities in Russian Military Thought, in Blank, *Russian Nuclear Weapons*, 265.

65  Interviewee: G.

66  McDermott, Russia's Conventional Armed Forces, 71.

67  Sokov, Nuclear Weapons in Russian Strategy, 198.

68  Arbatov and Dworkin, *Revising Nuclear Deterrence*, 8.

69 Sokov, Nikolai (2009) The Evolving Role of Nuclear Weapons in Russia's Security Policy, in Hansell, Christina and Potter, William C., eds., *Engaging China and Russia on Nuclear Disarmament* (Monterey, CA: James Martin Center for Nonproliferation Studies), 77.
70 Sokov, Why Do States Rely, 107.
71 Natural Resources Defence Council (2013) *From Mutual Assured Destruction to Mutual Assured Stability: Exploring a New Comprehensive Framework for U.S. and Russian Nuclear Arms Reductions* (Washington, DC: NRDC), 8, 35, 44, 67.
72 Sokov, Nikolai N., Yuan, Jing-dong, Potter, William C., and Hansell, Cristina (2009) Chinese and Russian Perspectives on Achieving Nuclear Zero, in Hansell and Potter, *Engaging China and Russia*, 13; Fatton, Lionel P. (2016) The Impotence of Conventional Arms Control: Why Do International Regimes Fail When they Are Most Needed? *Contemporary Security Policy*, 37(2), 214.
73 Postol, Theodore A. (2019) Russia may have Violated the INF Treaty. Here's How the United States Appears to have Done the Same, https://thebulletin.org, 14 February.
74 Kristensen, Hans M., McKinzie, Matthew and Postol, Theodore A. (2017) How US Nuclear Force Modernization Is Undermining Strategic Stability: The Burst-Height Compensating Super-Fuze, https://thebulletin.org/, 1 March.
75 Squassoni, Sharon (2017) Through a Fractured Looking-Glass: Trump's Nuclear Decisions so far, *Bulletin of the Atomic Scientists*, 73(6), 371.
76 Lonsdale, David J. (2019) The 2018 Nuclear Posture Review: A Return to Nuclear Warfighting? *Comparative Strategy*, 38(2), 105.
77 Gormley, Dennis M. (2015) US Advanced Conventional Systems and Conventional Prompt Global Strike Ambitions, *The Nonproliferation Review*, 22(2), 133.
78 Putin, Vladimir (2012), Being Strong: National Security Guarantees for Russia, http://archive.premier.gov.ru, 20 February.
79 Mehta, Rupal N. (2019) Extended Deterrence and Assurance in an Emerging Technology Environment, *Journal of Strategic Studies*, 19.
80 Woolf, Amy (2019) *Conventional Prompt Global Strike and Long-Range Ballistic Missiles: Background and Issues* (Washington, DC: Congressional Research Service), 46–47.
81 Horowitz, Michael C. (2019) When Speed Kills: Lethal Autonomous Weapon Systems, Deterrence and Stability, *Journal of Strategic Studies*, 42(6), 775.
82 Interviewee: D.
83 Tsypkin, Russian Politics, 784.
84 Interviewee: D.
85 Karaganov, Sergey (2011) Should We Overcome Deterrence? http://eng.globalaffairs.ru, 22 April; McDermott, Russia's Conventional Armed Forces, 79.
86 Sokov et al., Chinese and Russian Perspectives, 8.
87 Sokov, Evolving Role, 76.
88 Laruelle, Marlene (2010) *In the Name of the Nation: Nationalism and Politics in Contemporary Russia* (London: Palgrave Macmillan).
89 Tsypkin, Russian Politics, 782.
90 Arbatov et al., *Russia and Dilemmas*, 102.
91 Ibid.; Sokov, Chinese and Russian Perspectives, 7.
92 Arbatov, Russia,, 74.
93 Tsypkin, Russian Politics, 785.
94 Oxenstierna and Bergstrand, Defence Economics, 51.

95 Kagarlitsky, Boris (2002) *Russia under Yeltsin and Putin: Neo-liberal Autocracy* (London: Pluto), 6, 160.

96 Sumner, Daniel (2000) Russian Perceptions of Nuclear Weapons, www.acronym. org, 44, March; Podvig, Russian Public Opinion on Nuclear Weapons.

97 World Public Opinion, American and Russian Publics.

98 Interviewee: D.

99 Zlobin, Nikolai (2012) Putin's Beseiged Fortress, www.themoscowtimes.com, 16 February.

100 Tsypkin, Russian Politics, 793; Pew Research Center (2018), America's International Image Continues to Suffer, www.pewresearch.org, 1 October.

101 Tsypkin, Mikhail and Loukianova, Anya (2009) Formulation of Nuclear Policy in Moscow: Actors and Interests, in Hansell and Potter, *Engaging China and Russia*, 119.

102 Scowcroft, Brent, Hadley, Stephen J. and Miller, Franklin (2014) NATO-Based Nuclear Weapons Are an Advantage in a Dangerous World, www.scowcroft. com, 17 August; Tsypkin, Russian Politics, 794.

103 International Institute for Strategic Studies (2015) *The Military Balance 2015* (London: Routledge).

104 Carlsson, Marta and Norberg, Johan (2012) The Armed Forces, in Pallin, *Russian Military Capability*, 114; Rumer, *Russian Foreign Policy*, 74.

105 Natural Resources Defence Council (2013) *From Mutual Assured Destruction to Mutual Assured Stability Exploring a New Comprehensive Framework for U.S. and Russian Nuclear Arms Reductions* (Washington, DC: NRDC), 7.

106 Oxenstierna and Bergstrand, Defence Economics, 45.

107 OECD (2011) *Reviews of Labour Market and Social Policies: Russian Federation* (Paris: OECD).

108 Tsypkin and Loukianova, Formulation of Nuclear Policy, 118.

109 Rumer, *Russian Foreign Policy*, 51.

110 Kagarlitsky, *Russia under Yeltsin and Putin*, 135, 280.

111 Ibid., 280.

112 Anderson, Perry (2015) Unfittable Russia, *New Left Review*, July/August, 15.

113 Pavlovsky, Gleb (2014) Putin at Close Range, *New Left Review*, July/August, 65.

114 Seddon, Max (2020) 'Turbulence' Persuades Russia's Putin to Back Move to Extend his Rule, www.ft.com, 12 March.

115 Rumer, *Russian Foreign Policy*, 84.

# 5 United Kingdom

## 5.1 Introduction

This chapter assesses the explanatory power of *institutional democratisation,* alongside mainstream and realist approaches, regarding the causes and consequences of UK nuclear possession and disarmament. It proceeds over several sections, beginning by comparing and contrasting different perspectives on the UK's nuclear experience since WW2, before moving on to discuss modern-day British nuclear politics. The second section of this chapter summarises the mainstream and realist approaches explored in Chapter 2 in relation to the UK's particular experience as an NWS. This is principally done by placing the UK's development of nuclear weapons within the context of relevant historical events from the mid to late 20th and early 21st century—focusing, in particular, on the Cold War, when the UK became a NWS as part of a close military and political alliance with the US in opposition to the Soviet Union.

I also explore justifications for Britain retaining the bomb and arguments against disarmament from advocates of nuclear possession, including how the UK has officially justified its nuclear status over time. For example, the British government has historically presented nuclear possession as a necessity to ensure national security via deterrence given the external threats posed to the UK. Yet there are also strong domestic political factors driving the UK's nuclear weapons decision-making and international policy more generally, including those relating to the dynamics in and between political parties, the maintenance of elite groups' position in the social order, the stories told about Britain's role as a leading world power and bureaucratic continuity.

The third section of this chapter explores these domestic factors and considers the explanatory power of *institutional democratisation* regarding the causes and consequences of UK nuclear possession and disarmament. For example, official justifications for the UK's nuclear status are challenged and the limitations of the mainstream and realist literature's explanation of UK nuclear politics identified. This is done initially through an exploration of critical perspectives on the Cold War and the UK's global strategy to identify historical approaches that support the specific claims and ideas of *institutional democratisation.* Previous work focusing on the UK's domestic nuclear politics is also discussed here in order to specify the wider impact nuclear possession has had on the UK's polity, relate this to criticisms of the UK's record as a liberal democracy—and its maintenance of a sizeable and costly military establishment—and highlight existing scholarly arguments compatible with the main contentions of this study.

Having provided this historical overview of the different approaches to UK nuclear possession and disarmament, the fourth section of this chapter goes into more detail concerning current obstacles to and opportunities for *institutional democratisation* in the UK by reviewing modern-day UK politics in relation to the nuclear question. Evidence is also presented to show both how wide the democratic deficit is in the UK and how this is reflected in and relates to the gap between UK public opinion on nuclear arms control and disarmament and the actions of decision-makers. For example, the present state of the UK peace and disarmament movement and public opinion concerning the UK's role in the world is considered to explore civil society's potential contribution to disarmament initiatives and how these may develop and be strengthened as part of a wider democratisation process. Chapter 8 then contextualises the UK's responsibility for disarmament action alongside the other NWS.

## 5.2  Mainstream and realist perspectives on the causes and consequences of UK nuclear possession and disarmament

Whilst the UK was, following the US and USSR, the third nation to acquire nuclear weapons in 1952, it was also, Andrew Brown notes, 'the first state to take the decision to acquire an atom bomb'.[1] As Scott Sagan explains, realist explanations of the British decision to develop nuclear weapons, based on the security model, emphasise the 'growing Soviet military threat' and the reduced 'credibility' of US extended deterrence guarantees following the USSR developing the ability to 'threaten retaliation against the United States'.[2] British planners thus wanted to be able to deter the Soviet Union, for example, should the US 'go it alone' and start a war when they saw an advantage.[3] It was—and is—also believed that being a nuclear possessor allows the UK to retain influence in Washington, in order, as Roger Ruston vividly puts it, to have a 'say in the end of the world'.[4]

Margaret Gowing, meanwhile, identifies several factors driving the early British pursuit of nuclear weapons—in addition to its desire to be able to deter an 'atomically armed enemy', whether Nazi Germany or the Soviet Union. For example, as a great military power Britain must possess 'all major new weapons', with the bomb symbolising British strategic independence.[5] According to Andrew Dorman, 'successive British governments' have valued possessing an 'independent nuclear deterrent' because the experience of WW2 led to a recognition that 'ultimately one state might not be prepared to sustain massive losses for another'.[6] Furthermore, Prime Minister Winston Churchill justified the British bomb as being necessary to ensure that Soviet military targets 'would be given what we consider the necessary priority', based on what planners saw as the UK's unique vulnerability to nuclear attack owing to its size and proximity to Europe.[7]

In 1946 Foreign Secretary Ernest Bevin famously asserted about the new nuclear technology, despite its cost at a time of austerity, that the UK 'could not afford to acquiesce in an American monopoly of this new development', so that 'We've got to have the bloody Union Jack on top of it'.[8] As Malcolm Chalmers notes, another official British justification for nuclear possession was that the UK would act as a second centre of decision-making in Europe, meaning that the Soviet Union would more likely be 'deterred from aggression'.[9] In addition to deterring attacks on the British mainland, in the late 1950s the UK's airborne nuclear force was also tasked with a global role as part of the UK's commitments to NATO.[10]

Today, supporters of Britain retaining its Trident nuclear weapons system, such as former US Defense Secretary Ash Carter, argue that the UK can and should 'punch above its weight' in order to continue playing an 'outsized role' in the world.[11] According to this logic—found in the arguments of prominent pro-Trident figures, such as Commodore Tim Hare—by disarming, the UK would shirk its global responsibilities and would have to downgrade its ambitions, leading to a loss of influence and international status, as well as pressure to let go of its permanent seat on the UN Security Council. Disarmament would, moreover, cause the UK to lose the security and stability nuclear deterrence provides and could create the perception that the UK is a 'soft target for nuclear blackmail and intimidation' in the words of Peter Cannon.[12]

Such concerns, it is argued, are of crucial importance in today's volatile world. For example, Prime Minister Tony Blair remarked in 2006, prior to the then Labour government's decision on whether to replace the UK's Trident nuclear weapons system, that 'the United Kingdom should continue as a nuclear weapons state, for at least the next 50 years', because 'the one certain thing about our world today is its uncertainty'.[13] For adherents of this mindset, such as Lord George Robertson, recent events, including the Arab Spring, suggest that the future will be beset by continual surprises and shocks, some of a nuclear nature, heightening risk and the need for security guarantees. For example, North Korea has tested a nuclear weapon and Iran may be seeking to acquire one.[14]

Proponents of the UK remaining an NWS, such as Sir David Omand, also tend to aver that NATO's collective security system—including a framework of deterrence extended from the US to its European allies—is fundamental to the UK's national security.[15] Ending the UK's nuclear contribution would, Cannon therefore argues, weaken this system by making it dependent on France (as the only NWS in Europe) and the US.[16] Such a development would also, for Hare, likely open up new fault lines and stresses in the Euro-Atlantic relationship, jeopardising US commitments to Europe, something which Omand argues would not only affect British and European security but could also put at risk Washington's extended deterrence in the Far East and Pacific regions.[17]

In general, opponents of disarmament argue that the UK would therefore not gain anything from giving up its nuclear weapons, but that this would bring significant costs and risks. For example, Lord Robertson argues that such a move would not have any positive impact on global nuclear proliferation. This is, he argues, shown by the fact that the disarmament steps that the UK has already carried out have done nothing to discourage proliferation among those states that desire a nuclear weapons capability.[18] On the contrary, Conservative Way Forward propose that unilateral disarmament would ensure that the UK has no influence or leverage over multilateral disarmament negotiations.[19] Instead of moving to zero, the UK—for Lord Michael Boyce—should continue its work on developing verification technology and encourage the US and Russia to take the lead on disarmament, following on from the New START treaty.[20] Others highlight the economic costs of disarmament, including the jobs that would be lost. Whilst some senior British political figures have argued that economic and employment factors should not be a determining factor in deciding whether the UK remains an NWS, such concerns do play a prominent role in the debate.[21]

Opponents of disarmament, such as former Defence Secretary Sir Malcolm Rifkind, also point to the significant public support for the UK remaining a NWS.[22] In addition to supportive public opinion, the cross-party consensus between Conservative MPs and a majority of Labour MPs on maintaining the Trident system has recently been vital to the UK retaining its nuclear-armed status. Moreover, Prime Minister Blair and his supporters presented the government's approach to Trident renewal in 2006 and 2007 as being open and inclusive, with a 'strong and healthy debate', so that the subsequent parliamentary vote backing renewal could be framed as a democratic endorsement of government policy.[23]

In terms of a multilateral path to nuclear abolition, the UK government's position has been that it would join in such a process after the US and Russia have reached low numbers of nuclear weapons. For example, during Gordon Brown's tenure as Prime Minister the Foreign and Commonwealth Office produced an information paper titled *Lifting the Nuclear Shadow: Creating the Conditions for Abolishing Nuclear Weapons*, wherein the government outlined how it would fulfil its commitments under the NPT. The document stated that the UK would 'continue to work towards the total elimination of our own nuclear arsenal and all others through multilateral, mutual and verifiable agreements'. Furthermore, when 'useful', the government would willingly include in any negotiations 'the small proportion of the world's nuclear weapons that belong to the UK'.[24] Indeed, the UK has previously focused on the limited scale of its nuclear arsenal partly in order to help present itself as 'the most forward leaning, progressive and transparent nuclear weapon state in the P5'.[25]

## 5.3  Critical perspectives on the causes and consequences of UK nuclear possession and disarmament

As a critical perspective on the causes and consequences of UK nuclear possession and disarmament, *institutional democratisation* focuses on the role played by domestic politics. Such alternative approaches are required because defenders of the UK's nuclear status, such as Sir Michael Quinlan, hint at but do not explain the domestic 'political, institutional and other motivations, as distinct from security rationales' for, what he describes as, Britain's 'independent' possession of the bomb.[26] It is therefore necessary to look beyond orthodox analyses for evidence and ideas supporting *institutional democratisation* in relation to the UK case.

For example, according to John Simpson and Jenny Nielsen, rather than being subject to democratic influence and control, the original British decision to acquire nuclear weapons was taken by 'a small group of key cabinet members in private' and subsequent British governments 'continued to favour taking decisions through this process'.[27] Similarly, for Lawrence Freedman, the established British tradition in this area was one of 'secret and bipartisan policy-making' with an 'emphasis on continuity'.[28] Beatrice Heuser also observes that British political culture is less democratic than the US's, so that the normal secrecy on defence and nuclear issues is heightened in the UK and the public is 'told very little'. Heuser links this situation to the historic 'mutual fear' characterising the 'class divide' in the UK, which consists of the British establishment's 'fear of the masses', who elites see as an 'internal threat', whilst the people themselves are suspicious of 'the ruling classes'.[29]

The continuity which Freedman and Heuser refer to is explained, in relation to the bomb, by analysts such as Ritchie and Chalmers, who highlight the defence establishment's inherent conservatism, resistance to radical change and the bureaucratic and technological momentum that pushes British nuclear possession along.[30] For a nation of its size the UK's military establishment is particularly large, meaning that it carries significant weight in decision-making. As David Edgerton notes, this situation came about as, between 1939 and 1955, a 'military-industrial-scientific complex' was created that amounted to a 'warfare state', giving the UK a 'sharply differentiated third place in a bipolar world'.[31]

Despite the UK today being a declining power that is questioning its role in the world post-Brexit, the size and scale of the British 'warfare state' is still significant. The question of loss aversion is therefore particularly pertinent to discussions of the elimination of the British nuclear arsenal as this is seen by military-security elites as an irreversible move given the technological costs involved. The delicacy of this subject also explains why Blair, as Labour leader, took a pro-Trident stance and sought to unite the party behind this policy. This was seen as necessary so that the nuclear issue would not damage Labour's electoral prospects, as some within the party believed it had done in the 1980s when Labour supported unilateral disarmament, nor challenge the bipartisan consensus on defence matters and allow the Conservative party to use the issue as a political weapon, casting Labour as weak on security.

This phenomenon was seen in 2015 when then Prime Minister David Cameron alleged that Labour leader Jeremy Corbyn's ideas, including his long-standing support for unilateral nuclear disarmament, made his party a 'threat to national security'.[32] Labour's need to avoid giving the Conservatives opportunities to expose its internal divisions on this totemic issue likely contributed to the party maintaining its support for replacing Trident during Corbyn's tenure as leader. For Corbyn to have accomplished a swing within his party towards unilateralism would have been an impressive feat given that, in 2016, up to 130 Labour MPs then in Parliament supported the UK's possession of nuclear weapons, with up to 90 opposing.[33] Despite this, Corbyn himself kept open the question of a future Labour government's support for Trident and announced that he would not detonate nuclear weapons if he became Prime Minister.[34]

Turning to the question of the special relationship, the secrecy pervading the British nuclear weapon system during the Cold War became inextricably linked to Britain's near total dependence on the US over a range of areas. As Kristan Stoddart explains, Britain can only claim to have had any semblance of an independent nuclear weapons capability between 1952 and the mid-1960s.[35] In 1962 US President John F. Kennedy agreed to provide UK Prime Minister Harold Macmillian with the Polaris nuclear weapons system on terms that were seen as 'remarkably favourable' to Britain.[36] In exchange the UK committed to assign its nuclear force to NATO and target it in accordance with alliance plans.[37] As Freedman explains, without Polaris the UK would not have had 'any sort of credible nuclear capability', given the speed of the two superpower's technological development at that time.[38] The resulting 'co-operative nuclear alliance' led Sir Frank Cooper, the Permanent Under Secretary of Defence, to state in the mid-1980s that 'if you ask me whether the Americans have an undue degree of influence over British defence policy I would have to say yes'.[39] As Hugh Miall noted, in return for nuclear and intelligence material, Britain provided bases for its nuclear forces as well as diplomatic and military support when required—an arrangement that raised serious questions for British democracy and independence.[40]

Indeed, as Bruce Kent observes, it was Britain's subordination to the US that, in the early 1980s, ignited huge public opposition to nuclear weapons, following the government's decision to host US cruise missiles and replace Polaris with the far more powerful Trident nuclear weapons system.[41] The UK's vigorous anti-nuclear movement has existed since the early days of the bomb, with the Campaign for Nuclear Disarmament (CND) being founded in 1958. The influence of CND, and the wider British anti-war and peace movement's strength, has ebbed and flowed over the years, largely in response to the degree of public awareness and sense of threat regarding nuclear dangers. One recent survey shows that a majority of voters (54%) would prefer Britain to abandon its nuclear weapons and not replace them, whilst another survey shows that a larger majority (81%) favour an international plan 'for totally eliminating nuclear weapons according to a timeline'.[42] Thus, as a 2007 study by the Simons Foundation found, the UK 'boasts a high level of support for elimination of nuclear arms and nuclear testing all over the world'.[43]

As noted above, Blair's decision to hold a parliamentary vote on Trident replacement in 2007 may have given the impression that this process would reflect the breadth, depth and history of opposition to nuclear possession in the UK, and that it would be conducted in an accountable and open fashion, with democratic checks and balances. In reality, as Becker, Müller and Wisotzki note, Labour's earlier 1997/98 Strategic Defence Review resulted in an 'unequivocal commitment to nuclear weapons' with the following debate on the successor to Trident 'conducted within a small circle'.[44] Analysts note that Blair had thus already decided to replace Trident with a new system several years prior to the 2007 vote.[45]

Moreover, as Ian Davis argues, the procurement process for the UK's replacement of its nuclear-armed submarines had been altogether lacking in transparency. For example, despite parliament being asked to vote on whether to approve the construction of four new nuclear-armed submarines in 2016, work on the programme began in 2008, so that, for Davis, 'by the time of the anticipated review and vote at Main Gate it may already be too late to consider alternatives'.[46] The flaws in the UK's decision-making process were also shown in February 2020 when it was discovered that the UK had committed itself to buying a new generation of nuclear warheads, which will be based on US technology. Pentagon officials disclosed this decision before an official announcement was made by the British government.[47]

Despite the relatively low salience of nuclear issues for the British public today, there remains a backbone of researchers, parliamentarians and activists working to raise social consciousness regarding the opportunity costs and risks of nuclear weapons. This includes highlighting the safety and security problems at the UK's Atomic Weapons Establishment (AWE), and developing alternative policies such as defence diversification, to better imagine how disarmament could be managed and persuade sceptical communities and trade unions that jobs will be protected.[48] Such arguments appear to have made significant headway since, as Ritchie states, 'the British public appears quite firm in its support of global nuclear disarmament', whilst its support for the planned replacement of Trident is 'increasingly limited'.[49] Indeed, at a time of austerity the significant and rising sums dedicated to Trident drives opposition from several fronts, including civil society groups, the public, several senior political figures, such as Michael Portillo and David Owen, as well as high-ranking retired military officials.[50] Furthermore, the UK's 2010 National Security Strategy asserted that the UK does not currently face a 'major state military threat', opening up the questions of who and what Trident is meant to deter, why it is needed and who benefits from nuclear rearmament.[51]

Concerns and criticisms regarding the UK's undemocratic and secretive decision-making process on nuclear weapons complement arguments made by critics of the UK's overall political system. For example, David Beetham argues that Britain has become an 'unelected oligarchy' whereby the 'wealthy ... dominate public decision making'.[52] This state of affairs is the result of the 'dominance' of 'corporate and financial elites' over the government 'through the financing of political parties, think tanks and lobbying organisations, membership of advisory bodies, "revolving doors" and joint partnerships with government'.[53]

Such an assessment stands in stark contrast to evaluations provided by the Economist Intelligence Unit and Freedom House, which find that the UK scores highly across a range of indicators concerning democratic standards and civic rights.[54] It is notable, as previously mentioned, that none of these studies directly incorporate nuclear matters or military-related institutions into their methodology, possibly because such authors presume these issues to be recondite by nature, with relatively low political salience. Beetham, however, indicates the ways in which private commercial and industrial interests influence government decision-making in the UK by highlighting 'the links between the Ministry of Defence and the arms manufacturing companies, and the strong support given by government to their international trade along with other exporters'.[55]

## 5.4  UK nuclear politics today

Having provided an initial overview of how *institutional democratisation* adds explanatory value to our understanding of the causes and consequences of UK nuclear possession and disarmament, this section shall now look in more detail at contemporary British nuclear politics in order to provide a fuller analysis to support the main contentions of this study. Public discussion of whether the UK should remain an NWS grew as 2016 approached—which was when the final decision on whether to build a successor nuclear weapons system based on four submarines, to maintain continuous-at-sea-deterrence, was taken by Parliament. Yet, as noted above, analysts argued that approval for building the submarines was inevitable given that key political decisions had been made, contracts for essential materials agreed and work on upgrading infrastructure begun. Moreover, the impetus for these decisions can be largely explained by domestic factors, stemming from the overlapping needs and interests of several key economic, industrial and political actors and groups. In order to understand how the UK might eventually decide to relinquish its nuclear weapons I therefore explore in this section the politics of Trident replacement in terms of the strengths and weaknesses of the forces supportive of and opposing Britain remaining a nuclear power.

Of the scholars working on this issue, Nick Ritchie provides one of the most in-depth investigations of the forces working for the 'reproduction' of Trident in his *A Nuclear Weapons-Free World? Britain, Trident and the*

*Challenges Ahead.* In this work, Ritchie identifies four key 'enablers' of a like-for-like replacement for Trident. These include nuclear deterrence, national identity, the nuclear relationship with the US and the submarine-building industry. He then outlines the main 'resistances', namely, the impact on efforts to move towards global nuclear disarmament, the cost of the Trident replacement programme and alternative paths between a like-for-like replacement and unilateral nuclear disarmament.[56] Overall, Ritchie persuasively argues that it is the political benefits which nuclear weapons provide that allows them to continue occupying a central position in the self-image of the British 'political-defence establishment' as a 'major pivotal power', something which has become embedded over time and is difficult to dislodge.[57]

Indeed, as Tony Blair stated in his memoirs, Trident's utility in the post-Cold War world is 'less in terms of deterrence, and non-existent in terms of military use'. Blair admitted that what matter more to the military are 'helicopters, airlift and anti-terror equipment', yet giving up Trident would be 'too big a downgrading of our status as a nation', for which no Prime Minister wanted to take the blame.[58] What was being implicitly acknowledged here is that the UK's nuclear weapons are, and always have been, political weapons that operate on the domestic *and* international levels.

At home these weapons play an important role in helping pro-nuclear political leaders unify their parties, and the nation, behind the idea that the UK is a major global power. British elites also believe the UK to have global responsibilities as the 'closest ally' of the US, requiring a sizeable military budget for interoperable, expeditionary forces capable of intervening in crises and conflicts around the world.[59] This responsibility includes the UK playing its part in deterring and containing recalcitrant states such as Russia. Indeed, the threat posed by the Kremlin has been repeatedly utilised by pro-Trident commentators and political figures in recent years to justify increased military spending and the UK maintaining its nuclear status.[60] The power and appeal of this national narrative, which has its roots in Britain's victory in WW2 and resistance to the Soviet threat during the Cold War, means that nuclear possession continues to receive strong support amongst both Conservative and Labour voters.[61] Moreover, as Chalmers and Peter Hennessy explain, Trident is seen by political elites as 'essential in preserving Britain's position in Europe'. The need here is for the UK to be on a par with France and 'a notch above' non-nuclear Germany and Japan, which the UK lags behind in terms of economic strength.[62]

In terms of UK industrial concerns, Ritchie argues that ten companies, including defence giant BAE Systems, form a lobbying group for submarine procurement, whose main strength and weakness is that it is in a co-dependent relationship, where it is the only supplier and the Ministry of Defence (MOD) is the only customer.[63] Key to this situation is the fact that repeated government investigations—most recently the 2013 Trident Alternatives Review—have concluded that the optimal platform for the UK's nuclear weapons are ballistic missile submarines (SSBNs).[64] Moreover, without continual orders for submarines (including the Astute attack submarines which preceded the latest Dreadnought-class SSBNs) ensuring a regular 'drumbeat', the industry will wither away as core skills and experience go elsewhere, imposing a 'use it or lose it' imperative on the political executive.[65]

Regarding where the forces supportive of nuclear disarmament in Britain stand today, a survey of contemporary public opinion by BASIC found that the population 'remains deeply divided on nuclear weapons and choices around Trident replacement'.[66] According to the report 'polls suggest that opinion has moved towards relinquishing nuclear weapons after Trident when given a simple yes/no choice' and that opposition increases when people are made aware of the cost of replacing Trident.[67] The recent referendum concerning Scottish independence also highlighted the significant public opposition to Trident north of the border.[68] The ruling Scottish National Party (SNP) pledged to make nuclear weapons illegal and force their removal from the Clyde naval base, which some, such as Vice-Admiral John McAnally, believe would lead to Britain being forced to abandon its nuclear weapons for good, principally owing to the cost of relocating Trident.[69] The SNP's case for removing Trident from Scottish territory and waters also focuses on opportunity costs and the fact that these weapons symbolise the 'democratic deficit' whereby Westminster imposes its policies on Scotland.[70]

Current spending projections for Trident replacement show that nuclear weapons will eat up a third of the MOD's overall annual equipment budget for about fifteen years from the early 2020s.[71] These budget cuts have led senior military figures, particularly from the Army, to express scepticism regarding the necessity of large-scale defence projects, which analysts such as Richard Norton-Taylor have interpreted as including Trident.[72] Construction of the first of four new SSBNs to carry the UK's nuclear weapons began in October 2016, yet a 2017 review of the project by the government's Infrastructure and Projects Authority found that 'successful delivery of the project is in doubt, with major risks or issues apparent in a number of key areas'.[73]

Disarmament advocates, such as the Nuclear Information Service, focus on the escalating costs and rising risks of the UK's nuclear weapons programme to explain the opportunity costs of nuclear possession and make the issue visible and relevant to people's lives.[74] Campaign groups such as CND and Global Zero also highlight how many schools and hospitals could be built if governments decided not to waste money on nuclear weapons.[75] Such arguments have gained weight following the publication of recent estimates suggesting that the whole-life cost of the UK's next generation of nuclear weapons may be over £200 billion, although it could be argued that even this figure does not reflect the wider costs of the UK remaining a NWS with Trident, given the need to maintain a viable submarine industry.[76] Elsewhere, a variety of prominent UK and US-based analysts have drawn attention to the possibility that submarines will in future be much more vulnerable to cyber-attacks and detection by underwater technologies and that ballistic missiles could even become obsolete.[77] This discussion has again led to questions being raised concerning the wisdom of the UK investing so many resources in building four new SSBNs given that it solely relies on this platform to deliver its nuclear weapons.

Another episode illustrating the political complexities of the British nuclear debate occurred in 2012 when the SNP, having promised to remove Trident from Scotland upon gaining independence, committed itself to remaining in NATO. This raised the difficult question of how Scotland could reject Trident on the one hand whilst accepting membership of an alliance which does not rule out a nuclear first strike.[78] The issue of how a non-nuclear Britain would relate to a nuclear-armed NATO was also a vexing issue for the Labour party under Neil Kinnock, which was, until 1989, committed to unilateral nuclear disarmament, withdrawal from NATO and the removal of US bases from the British Isles. The contradictions within Labour's position could not be sustained and led the party to replace unilateralism with multilateralism, whereby the UK would only place its nuclear weapons in disarmament negotiations once the superpowers were ready to commit to abolition. The significance of this move was that Labour was not prepared to move towards the kind of neutralism that E. P. Thompson understood nuclear disarmament to entail when he wrote of the need to end British subservience, by 'shaking off' superpower hegemony to 'reclaim autonomy'.[79]

As it is, the UK remains one of, if not the, most ardent promoters of NATO, with most 'opinion-formers' surveyed seeing the alliance as either a 'vital' or 'important' pillar of security for the UK, according to a 2015 report produced by Chatham House.[80] Domestically, whilst vocal critics of the alliance exist within the Labour and Scottish National parties, for the most part it retains strong parliamentary support. This is despite evidence showing considerable public disquiet with the status quo, including a 2015 YouGov poll which found that a majority of swing Labour voters want the party to be 'less subservient to the USA', not 'get involved in American wars' and instead be 'more positive about Britain's role in Europe'.[81] In response, then Labour leader Jeremy Corbyn argued that the alliance needs to be brought 'under democratic control' and consider carefully future eastwards expansion.[82]

The impact of the UK leaving the European Union (EU) is likely to push the UK closer to the US and thus NATO, as it uses its military, intelligence and security capabilities as leverage to maintain political influence on the continent. For Becker, Müller and Wisotzki, Britain's membership of the EU has, in the past, had some impact on the UK government's nuclear policy, motivating it to make some compromises with the 'more proactive fellow EU members', for example, 'transparency or the reduction of tactical nuclear weapons' in relation to EU Common Positions at NPT Review Conferences.[83] Without such diplomatic constraints, however limited, the UK is even more unlikely, at a time of national turbulence, to voluntarily divest itself of such a symbol of political power.

It is worth noting that in an article reflecting on the UK government's addiction to nuclear weapons, George Monbiot argues that the one force that could finally 'kill' Trident is the US. For only once the US has begun to dismantle its huge nuclear arsenal and 'ordered' the UK to follow suit would such disarmament occur. Recalcitrant parliamentary and public opinion in the UK thus 'counts for nothing'.[84] An alternative to waiting for orders from the US is for the UK to, as Mark Curtis argues, in a similar vein to Thompson, 'withdraw its general backing for Washington and instead pursue a policy of strategic non-cooperation'. Such a seismic shift, replacing 'a very well-entrenched, elitist, secretive and totalitarian domestic governance system' with a 'genuine popular democracy' will, for Curtis, only result from 'massive public pressure'.[85]

In terms of the potential for a shift in politics and public opinion that would lead to such a transformation in the UK's international policy, a 2014 YouGov poll found that a majority of the public thinks that the UK should aspire to be a 'great power' rather than accept that it is in decline; 61% of respondents also thought that NATO is either 'vital' or 'important' to UK security.[86] Moreover, a larger number of respondents thought that the UK's closest ties should be with the EU (30%) rather than the US (25%), suggesting the European dimension of the alliance may be more important to the public.[87] Overall, such findings suggest that the first job for those seeking to develop an alternative to Britain's current nuclear policy is that they lead a public debate about what type of international behaviour and relationships would align best with the values and goals that voters most care about.

Elsewhere, the individuals and groups working to increase public awareness of the economic, environmental and political costs and risks of the UK remaining an NWS and interventionist military power support *institutional democratisation* by opening up British defence and foreign policy to some degree of transparency and accountability. Yet whilst progress has been made, several of the proposals put forward in the 1980s by civil society to increase access to information and public and parliamentary control over nuclear weapons have still yet to be realised.[88] Such democratic accountability is essential if the requisite pressure for nuclear disarmament and a sustainable approach to security is to be built over the long term in the UK.

## 5.5 Conclusion

This chapter has sought to assess the explanatory power of both the mainstream and realist and the *institutional democratisation* perspectives on the causes and consequences of British nuclear possession and disarmament. For the UK case study, the security model is somewhat valuable in explaining the wartime origins of the British decision to acquire the bomb, but cannot satisfactorily account for the domestic and international factors shaping the UK's nuclear politics, past and present. For example, the security model focuses on external threats—particularly emanating from Russia—as driving the UK's continued need for nuclear deterrence.

Yet in recent years the UK government has stated that it does not perceive a military threat to the UK from a major state, preferring to argue that future uncertainty necessitates maintaining the bomb as the ultimate insurance policy. Moreover, the UK defence and foreign policy establishment highly values its nuclear status because it believes this elevates it beyond being another middle ranking power as well as offering moral and political support to the US in its management of international order. However, being such a close ally of the US brings with it its own dangers, so that London's propensity to support Washington's military interventions and NATO expansion may generate a scenario whereby British territory or vital interests are threatened by nuclear attack—which is a strong but unspoken reason why British defence and foreign policy-making elites continue to believe that the UK must persist in practising nuclear deterrence.

The security model also has little to say about the important domestic factors driving the UK's continuing nuclear possession and the changes that need to take place within the UK for nuclear disarmament to become politically viable. *Institutional democratisation* does a better job in both these areas by highlighting: the closed and secretive nature of UK nuclear weapons decision-making; how this connects to the UK's sizeable 'warfare state' and the flaws within the UK's political system; and the impact that nuclear possession has had on the British polity and political discourse. As a theory of political change, *institutional democratisation* also helps us imagine a path by which the UK could shift its international policy away from power projection and nuclear possession and towards a security policy consisting of a restrained posture, utilising primarily defensive, conventional capabilities whilst prioritising political over military solutions to regional and global problems.

By studying the domestic political scene, including industrial, economic and party-political factors, as well as public opinion, it is thus possible to propose how UK politics might develop in future in ways that support nuclear disarmament action. For example, civil society has fought for decades for information regarding the UK's nuclear arsenal to be in the public domain so that its costs and risks, including safety and environmental issues surrounding AWE, are widely known. Maintaining work of this kind, as well as building social movements advancing alternative ideas, including an industrial policy focused on defence diversification and an international policy focused on common global security, could develop existing public enthusiasm for nuclear disarmament and draw in new supporters.

Whilst it is clear that there are significant barriers to *institutional democratisation* in the UK, including the grip the British guardianship has over nuclear weapons decision-making and the UK's subordinate relationship to the US—particularly regarding international policy—the potential exists for civil society to deepen British democracy and transform the UK's international posture in the short to medium term as a means to advance the cause of nuclear disarmament. Clearly, such a prospect frightens powerful supporters of the status quo, who seek to paint their opponents as threats to national security. Yet large sections of the British public favour scrapping nuclear weapons and the prospect of Trident's removal from Scotland remains a possibility as long as support for independence remains strong.

What should also be clear from the preceding discussion is that it will likely take a significant shift in international politics, principally concerning the UK–US relationship, if British military and political elites are to abandon nuclear deterrence, as proposed by the *guardianship* approach to nuclear arms control and disarmament. Such elites are so tightly bound to the bomb by history, identity and established structures of power, including the nuclear relationships with France and the US, that the costs of relinquishing the political power vested in Trident are too great for them to contemplate, whilst the political benefits at home and abroad are not sufficiently attractive. At the same time, the UK has come closest of the NWS to unilateral disarmament, with some Labour party leaders clearly more comfortable with this notion than others. It is also possible that significant cost escalation for the next generation SSBN programme, industrial troubles and technological developments (including in cyber and underwater warfare) may combine to make Trident ineffective and vulnerable from both a political and military viewpoint, so that alternative policy options—including disarmament—become more inviting over time.

Leaving such technocratic scenarios aside, British nuclear disarmament would likely only come about as a result of two different developments, which may be complementary depending on how they emerge. First, a disarmament initiative led by the US, for example on a multilateral basis, involving Russia and with Chinese and French support, would very likely lead to British participation. Secondly, if a popular domestic movement emerges which is committed to democratising British institutions, is capable of challenging the power of defence and foreign policy elites and is able to win control over nuclear weapons decision-making, the possibility of irreversible disarmament measures being enacted becomes feasible. Such a disarmament process would also require an alternate vision for Britain's role in the world, so that collective and non-military approaches to security—with a focus on human rights, as well as conflict prevention and resolution—are prioritised and international law is adhered to regarding the threat or use of force.

## Notes

1 Brown, Andrew (2008) Historic Barriers to Anglo-American Cooperation, in Mack, Jenifer and Cornish, Paul, eds., *US-UK Nuclear Cooperation After 50 Years* (Washington DC: CSIS), 37.
2 Sagan, Scott D. (1996) Why Do States Build Nuclear Weapons? Three Models in Search of a Bomb, *International Security*, 21(3), Winter, 58.

3 Coleman, David G. and Siracusa, Joseph M. (2006) *Real-World Nuclear Deterrence: The Making of International Strategy* (Westport, CT: Praeger), 25.

4 Ruston, Roger (1989) *A Say in the End of the World: Morals and British Nuclear Weapons Policy 1941–1987* (Oxford: Clarendon Press).

5 Gowing, Margaret (1974) *Independence and Deterrence*, vol. 1, *Britain and Atomic Energy, 1945–52: Policy Making* (London: Palgrave Macmillan), 184–185.

6 Dorman, Andrew (2017) *The Future of British Defence Policy*, www.ifri.org, July, 15–16.

7 Clark, Ian and Wheeler, Nicholas (1989) *The British Origins of Nuclear Strategy: 1945–1955* (Oxford: Clarendon Press), 10–12, 71.

8 Brown, Historic Barriers, 39.

9 Chalmers, Malcolm (1984) *Trident: Britain's Independent Arms Race* (London: CND), 11.

10 Stoddart, Kristan (2012) *Losing an Empire and Finding a Role: Britain, the USA, NATO and Nuclear Weapons, 1964–70* (Basingstoke: Palgrave Macmillan), 80.

11 Reuters (2016) U.S. Defence Secretary Says UK Needs Nuclear Weapons for 'Outsized' World Role, www.reuters.com, 13 February.

12 Hare, Tim (2011) Evidence Submitted by Commodore Tim Hare, *BASIC Trident Commission*, www.basicint.org; Cannon, Peter (2012) The Necessity of Nuclear Deterrence, *BASIC Trident Commission*, www.basicint.org, 7.

13 Booth, Ken and Wheeler, Nicholas J. (2008) *The Security Dilemma: Fear, Cooperation and Trust in World Politics* (Basingstoke: Palgrave), 266.

14 Robertson, George (2011) Evidence to BASIC Trident Commission, *BASIC Trident Commission*, www.basicint.org.

15 Omand, David (2011) Evidence to BASIC Trident Commission, *BASIC Trident Commission*, www.basicint.org. 1

16 Cannon, Necessity of Nuclear Deterrence, 3.

17 Hare, Evidence, 3; Omand, Evidence, 2.

18 Robertson, Evidence, 1.

19 Conservative Way Forward (2012) Submission of Written Evidence to the BASIC Trident Commission, *BASIC Trident Commission*, www.basicint.org, 3.

20 Boyce, Michael (2011) BASIC Commission: Future of UK Nuclear Weapons Policy, *BASIC Trident Commission*, www.basicint.org, 3.

21 Rifkind, Malcolm, Browne, Des, Campbell, Menzies, Bailes, Alyson, Greenstock, Jeremy, Guthrie, Charles, Hennessy, Peter and Rees, Martin (2014) *The Trident Commission, Concluding Report*, https://basicint.org, 6.

22 Rifkind, Malcolm (2014) *A Conservative Approach to the Forthcoming Debate on Trident* (London: BASIC/WMD Awareness), 7.

23 BBC (2006) Blair's Trident Statement in Full, http://news.bbc.co.uk, 4 December; BBC (2007) Trident Plan Wins Commons Support, http://news.bbc.co.uk, 15 March.

24 UK FCO (2009) *Lifting the Nuclear Shadow: Creating the Conditions for Abolishing Nuclear Weapons*. Accessed at: https://carnegieendowment.org/2009/02/05/lifting-nuclear-shadow-creating-conditions-for-abolishing-nuclear-weapons-pub-22715, 12.

25 Harvey, Nick (2011), UK-Norway Nuclear Disarmament Verification Initiative Workshop, www.gov.uk, 7 December.

26 Quinlan, Michael (2009) *Thinking about Nuclear Weapons: Principles, Problems, Prospects* (Oxford: Oxford University Press), 121.

27  Simpson, John and Nielsen, Jenny (2010) The United Kingdom, in Born, Hans, Gill, Bates and Hänggi, Heiner, eds., *Governing the Bomb: Civilian Control and Democratic Accountability of Nuclear Weapons* (Oxford: Oxford University Press), 86.

28  Freedman, Lawrence (1980) *Britain and Nuclear Weapons* (London: Macmillan), 54.

29  Heuser, Beatrice (1998) *Nuclear Mentalities? Strategies and Beliefs in Britain, France and the FRG* (London: Palgrave Macmillan), 16–21.

30  Ritchie, Nick (2006) *Replacing Trident: Who will Make the Decisions and How?* (London: Oxford Research Group); Chalmers, Malcolm (2010) *Nuclear Narratives: Reflections on Declaratory Policy* (London: RUSI), 2.

31  Edgerton, David (2006) *Warfare State: Britain, 1920–1970* (Cambridge: Cambridge University Press), 1–3.

32  Stone, Jon (2015) David Cameron Claims Jeremy Corbyn Is a 'Threat to National Security', www.independent.co.uk, 13 September.

33  Riley-Smith, Ben (2016) 130 Labour MPs to Back Trident Despite Jeremy Corbyn's Opposition, Anti-Nuclear Campaigners Predict, www.telegraph.co.uk, 30 January.

34  Wintour, Patrick (2015) Jeremy Corbyn: I Would Never Use Nuclear Weapons if I Were PM, www.theguardian.com, 30 September.

35  Stoddart, *Losing an Empire*, 14.

36  Freedman, *Britain and Nuclear Weapons*, 25.

37  Davis, Ian (2015) *The British Bomb and NATO: Six Decades of Contributing to NATO's Strategic Nuclear Deterrent* (Stockholm: SIPRI), 20.

38  Freedman, *Britain and Nuclear Weapons*, 8.

39  Stoddart, *Losing an Empire*, 14; Miall, Hugh (1987) *Nuclear Weapons: Who's in Charge?* (Basingstoke: Macmillan), 77.

40  Ibid.

41  Kent, Bruce (2003) Resisting the British Bomb: The 1980s, in, Holdstock, Douglas and Barnaby, Frank, eds., *The British Nuclear Weapons Programme 1952–2002* (London: Frank Cass), 64.

42  World Public Opinion (2008) Publics around the World Favor International Agreement to Eliminate All Nuclear Weapons, www.worldpublicopinion.org, 9 December; Glover, Julian (2009) Voters Want Britain to Scrap All Nuclear Weapons, Icm Poll Shows, www.theguardian.com, 13 July.

43  Simons Foundation (2007) *Global Public Opinion on Nuclear Weapons* (Vancouver: Simons Foundation), September.

44  Becker, Una, Müller, Harald and Wisotzki, Simone (2008) Democracy and Nuclear Arms Control—Destiny or Ambiguity? *Security Studies*, 17(4), 831–832.

45  Ripley, Tim (2004) (Not-So) Secret Plans for Trident Replacement, www.scotsman.com, 9 June; Brown, Colin (2005) Revealed: Blair to Upgrade Britain's Nuclear Weapons, www.independent.co.uk, 2 May.

46  Davis, Ian (2011) *Trident 'Delay': Are we being Fooled Again?* UK Parliamentary Briefing (London: Greenpeace), January, 3.

47  Doward, Jamie (2020) Pentagon Reveals Deal with Britain to Replace Trident, www.theguardian.com, 22 February.

48  Burt, Peter (2017) *Playing with Fire: Nuclear Weapons Incidents and Accidents in the UK*, www.nuclearinfo.org; Burt, Peter (2016) *AWE: Britain's Nuclear Weapons Factory*, www.nuclearinfo.org.

49  Ritchie, Nick (2012) *A Nuclear Weapons-Free World? Britain, Trident and the Challenges Ahead* (Basingstoke, Palgrave Macmillan), 153.

50  Norton-Taylor, Richard (2009) Trident Replacement Plan No Longer Credible, Says Former Foreign Secretary, www.theguardian.com, 25 October; BBC (2012) Michael Portillo on Trident Nuclear Replacement Plans, www.bbc.co.uk, 2 November.

51  HM Government (2010) *A Strong Britain in an Age of Uncertainty: The National Security Strategy*, https://assets.publishing.service.gov.uk, 14.

52  Beetham, David (2011) *Unelected Oligarchy: Corporate and Financial Dominance in Britain's Democracy* (Liverpool: Democratic Audit), 11.

53  Ibid., 2.

54  Economist Intelligence Unit (2016) *Democracy Index 2016: Revenge of the 'Deplorables'*, ww.eiu.com, 7; Freedom House (2017) *Populists and Autocrats: The Dual Threat to Global Democracy*, https://freedomhouse.org, 24.

55  Beetham, *Unelected Oligarchy*, 19–20.

56  Ritchie, *Nuclear Weapons-Free World*, 51.

57  Ibid., 4, 77.

58  Blair, Tony (2007) *A Journey* (London: Random House), 636.

59  Ritchie, *Nuclear Weapons-Free World*, 77.

60  Clark, David (2016) *The Left's Nuclear Choice*, https://fabians.org.uk, 7 January; Bradley, Charlie (2019) Military Cuts Put Putin on War Path, www.express.co.uk, 15 December.

61  Grice, Andrew (2016) Trident: Majority of Britons Back Keeping Nuclear Weapons Programme, Poll Shows, www.independent.co.uk, 24 January.

62  Chalmers, *Trident: Britain's Independent Arms Race*, 52; Hennessy, Peter (2010) *The Secret State: Preparing for the Worst, 1945–2010* (London: Penguin), 79.

63  Ritchie, *Nuclear Weapons-Free World*, 107.

64  HM Government (2013) *Trident Alternatives Review*, www.gov.uk.

65  Ritchie, *Nuclear Weapons-Free World*, 114.

66  Ritchie, Nick and Ingram, Paul (2013) *Trident in UK Politics and Public Opinion* (London: BASIC), 1.

67  Ibid.

68  What Scotland Thinks (2014) Opinion Polls, www.whatscotlandthinks.org.

69  Johnson, Simon (2012) Independent Scotland would Ban Nuclear Weapons But Join Nato, www.telegraph.co.uk, 7 October; McAnally, John (2014) Scotland will be Powerless to Defend itself, www.telegraph.co.uk, 16 March.

70  Scottish National Party (2013) SNP Condemns UK Gov Trident Obsession, www.snp.org, 15 December.

71  Chalmers, Malcolm (2013) *Mid-Term Blues? Defence and the 2013 Spending Review* (London: RUSI), 12.

72  Norton-Taylor, Richard (2013) British Defence Chief Fires Volley of Warning Shots, www.theguardian.com, 19 December.

73  UK Infrastructure and Projects Authority (2017) *Annual Report on Major Projects 2016–17*, https://assets.publishing.service.gov.uk, 16, 19.

74  Cullen, David (2019) *Trouble Ahead: Risks and Rising Costs in the UK Nuclear Weapons Programme*, www.nuclearinfo.org.

75  CND (2014) *No to Trident*, www.cnduk.org; Global Zero (2014) *Cut Nukes*, cutnukes.globalzero.org.

76  CND (2016) *Trident Replacement Cost Rises to £205 Billion*, www.cnduk.org; Piper, Elizabeth (2015) Exclusive: Trident Programme to Cost 167 Billion Pounds, Far More than Expected, www.uk.reuters.com, 25 October; Street, Tim (2016) *Labour's Defence Policy Review Written Submission by Oxford Research Group Part Two: The UK's Nuclear Future*, www.oxfordresearchgroup.org.uk, April.

77  Chalmers, Malcolm (2013) Towards the UK's Nuclear Century, *RUSI Journal*, 158(6), 18–28; Futter, Andrew (2016) Is Trident Safe from Cyber Attack?, www.europeanleadershipnetwork.org, February; Hambling, David (2016) The Inescapable Net: Unmanned Systems in Anti-Submarine Warfare, www.basicint. org, 29 February; Parry, Chris (2015) Labour should Keep its Powder Dry on Trident: The Debate has Barely Begun, www.guardian.com, 24 November; Watt, Nicholas (2015) Trident could be Vulnerable to Cyber-Attack, Former Defence Secretary Says, www.guardian.com, 24 November; Clark, Bryan (2015) *The Emerging Era in Undersea Warfare* (Washington, DC: CSBA).

78  Carrell, Severin (2012) Alex Salmond Gains Slim SNP Vote for Joining Nato, www.theguardian.com, 19 October; Whitaker, Andrew (2014) Scottish Independence: 'American-Led Wars Risk', www.scotsman.com, 24 January.

79  Thompson, E. P. (1981) *Beyond the Cold War* (London: Merlin Press), 27; Thompson, E. P. (1987) How Britain Could Break the Ice in the Cold War, *Independent*, 25 February.

80  Raines, Thomas (2015) *Internationalism or Isolationism? The Chatham House–YouGov Survey British Attitudes towards the UK's International Priorities* (London: Chatham House), January, 26.

81  YouGov (2015) *Analysis: Should Labour Abandon the Centre Ground?* https:// yougov.co.uk, 29 January.

82  Simons, Ned (2015) Nato Should Have Been Disbanded in 1990, Says Jeremy Corbyn, www.huffingtonpost.co.uk, 27 August.

83  Becker et al., Democracy and Nuclear Arms Control, 834.

84  Monbiot, George (2010) War with the Ghosts, www.monbiot.com, 23 March.

85  Curtis, Mark (2003) *Web of Deceit: Britain's Real Role in the World* (London: Vintage), 436.

86  Raines, *Internationalism or Isolationism*, 26.

87  Ibid., 3.

88  Elworthy, Scilla (1989) Nuclear Weapons Decision-Making and Accountability, in Marsh, Catherine and Fraser, Colin, eds., *Public Opinion and Nuclear Weapons* (Basingstoke: Macmillan), 175; Miall, *Nuclear Weapons*, 157.

# 6 France

## 6.1 Introduction

This chapter assesses the explanatory power of *institutional democratisation,* alongside mainstream and realist approaches, regarding the causes and consequences of French nuclear possession and disarmament. As with the British case study, the chapter proceeds over several sections, beginning by comparing and contrasting different historical views on France's nuclear experience since WW2, before moving on to discuss modern-day French nuclear politics. Whilst there is a significant range of data covering French nuclear politics available in English, there are fewer sources than on UK and US nuclear politics for several reasons. These include the language barrier, the limited size of the audience for this subject beyond France, and the fact that for several decades there has been a national French pro-bomb consensus—though questioning of France's nuclear status has become more frequent in recent years. For these reasons, as well as the fact that the French nuclear arsenal is much smaller than that of Russia and the US, and the comparatively limited nature of France's power projection capabilities—which means that France does not play as great a role in other state's strategic calculus—this chapter is shorter than those on the UK, US and Russia. Chapter 8 will contextualise France's responsibility for disarmament action alongside the other NWS.

Section 6.2 summarises the mainstream and realist approaches explored in Chapter 2 in relation to France's particular experience as an NWS and assesses the merits of the security model as applied to France. This is principally done by placing the French development of nuclear weapons within the context of relevant historical events from the mid to late 20th and early 21st century—focusing, in particular, on the Cold War, where France positioned itself as a world power capable of deterring any threat to its national independence and security by virtue of its nuclear status. I also explore other justifications for France retaining the bomb and arguments against disarmament made by prominent advocates of nuclear possession in relation to relevant views on the origins and meaning of the Cold War.

Section 6.3 then assesses the explanatory power of *institutional democratisation* regarding the causes and consequences of French nuclear possession and disarmament by developing the domestic politics model. In doing so, official justifications for France's nuclear status are challenged and the limitations in the mainstream and realist literature's explanation of French nuclear politics identified. A brief exploration of critical perspectives on the Cold War and French global strategy is also presented to identify historical approaches that support the specific claims and ideas of *institutional democratisation.* Previous work focusing on France's domestic nuclear politics is discussed here in order to specify the wider impact nuclear possession has had on the French polity, relate this to criticisms of France's record as a liberal democracy and its maintenance of a sizeable and costly military establishment, and highlight existing scholarly arguments compatible with the main contentions of this study. As we shall see, the French case is particularly relevant to the claims made by *institutional democratisation* given the existence of what several scholars refer to as France's 'nuclear monarchy', which was initiated by President Charles de Gaulle and which has shaped the nature of the French state and Presidency to this day.

After this historical overview of the different approaches to French nuclear possession and disarmament, section 6.4 goes into more detail concerning current obstacles to and opportunities for *institutional democratisation* in France by reviewing modern-day French politics in relation to the nuclear question. A range of evidence is presented to show both how wide the democratic deficit is in France and the extent to which this is reflected in and relates to French public opinion on nuclear arms control and disarmament, and the actions of the French government. For example, France's political establishment maintains an apparent consensus on nuclear policy that makes the prospects for French nuclear disarmament appear distant, particularly when compared to the UK.

I subsequently consider how France can move beyond the status quo and advance nuclear disarmament, nationally and internationally. For example, the present state of the French peace and disarmament movement and public opinion concerning France's role in the world is considered to explore civil society's potential contribution to disarmament initiatives and how these may develop and be strengthened as part of a wider democratisation process. Potentially conducive developments in international security that will support such domestic political change highlighted in the expert literature are also discussed, to illustrate how domestic and international nuclear politics interact in the French case.

## 6.2 Mainstream and realist perspectives on the causes and consequences of French nuclear possession and disarmament

Realist explanations of the French decision to develop nuclear weapons, for Scott Sagan, are 'very simple' and emphasise the military threat posed by the USSR from the 1950s onwards as well as the reduced 'credibility' of US extended deterrence guarantees following the USSR developing the ability to 'threaten retaliation against the United States'.[1] As Beatrice Heuser notes, President Charles de Gaulle is often lauded for 'presiding over France's acquisition of nuclear weapons', which restored French sovereignty, independence and freedom after the Nazi occupation in WW2.[2] According to the security model, France's desire to keep hold of the bomb, Sagan adds, was then reinforced by the experience of Suez in 1956 when Paris had to 'withdraw its military intervention forces after a nuclear threat from Russia and under U.S. economic pressure'.[3] By becoming a nuclear power, and a leader in global affairs once more, France would ensure the national weakness displayed during this time was never repeated.

Wilfrid Kohl highlights other reasons why influential groups within the Fourth Republic—technocrats, military officers and politicians—pushed for the development of the bomb, ranging from:

> France's declining influence in NATO and the frustrations caused by the loss of her colonial territories to a desire for the modern weapons for the French army to restore its morale, to offset the effects of German rearmament, and to diminish France's dependence upon American military protection.[4]

France would thus be able to challenge US 'hegemony', rising to a status equal to the UK and above West Germany.[5] The French leadership role in Western Europe would also be retained, whilst it strove to unify the European continent 'from the Atlantic to the Urals', promote détente with Communist countries and secure an independent position for Europe in world politics.[6]

Supporters of France's nuclear status have been able to resist calls for disarmament at home and abroad, observers such as Oliver Debouzy and Michael Harrison argue, because whilst the French left opposed the Gaullist vision and supported nuclear disarmament in the 1960s, since the 1970s there has been a broad national consensus in favour of France's security and defence policies, including the retention of nuclear arms.[7] Heuser therefore usefully points out that nuclear enthusiasts argue, in response to claims that the French President's 'monopoly control' of the bomb was 'illegitimate because anti-democratic', that the President is authorised to use nuclear weapons as an enactment of the popular will in national elections.[8]

Following the end of the Cold War, Yost states that the French reconfigured their nuclear forces to deter two different threats, posed by: i) the re-emergence of a major state (potentially China or Russia), ii) a regional power armed with WMD.[9] The focus on the latter led President Jacques Chirac to emphasise the need to destroy the 'political, military and economic power centres' of an aggressor, requiring France to acquire, as Yost puts it, a wider range of nuclear options 'including more precise and more discriminate strike capabilities' which 'explicitly' lowered the threshold for use.[10] Elsewhere, *Libération* explained that the aim of these new capabilities was to be able to 'decapitate' a regime, 'without killing millions of innocent civilians'.[11]

The French 2013 Defence White Paper described the nation's nuclear weapons policy as follows:

> Our deterrence capability is strictly defensive. The use of nuclear weapons would only be conceivable in extreme circumstances of legitimate self-defence. In this respect, nuclear deterrence is the ultimate guarantee of the security, protection and independence of the Nation.[12]

As Yost explains, the concept of 'non-use' provides the French with political advantages, such as being able to remain in NATO and yet independent from it, and the US, in terms of nuclear deterrence strategy. France has thus never been a member of NATO's nuclear planning group.[13] In recent times the French bomb has also been presented as necessary to protect the nation's 'vital interests', although as Yost points out, what these mean 'depends on the President'.[14] For example, President Chirac outlined his belief that 'the integrity of our territory, the protection of our population' and 'the free exercise of our sovereignty' were at the core of France's vital interests. He later added that 'safeguarding our strategic supplies' and 'the defence of allied countries' could be invoked in the event of 'an unbearable act of aggression, threat or blackmail perpetrated against these interests'.[15]

In terms of arms control and disarmament, the French government claims that it has a 'unique, exemplary record in nuclear disarmament' and has taken 'significant unilateral steps' to abide by its NPT commitments, both in terms of nuclear and general and complete disarmament.[16] As proof of this, French officials cite several actions, including President Sarkozy's announcement in 2008 of 'a reduction by one third in the number of nuclear weapons, missiles and aircraft of the French airborne component' and the 2008 disarmament action plan it presented with European nations, endorsed by 27 EU Heads of State and Government.[17]

Notwithstanding such official claims regarding its record on arms control and disarmament, France's 2017 Defence and National Security Strategic Review stated that the nation's nuclear weapons were needed 'more than ever' and would be maintained over the 'long-term' because the 'nuclear factor is set to play an increasing role in France's strategic environment'.[18] Given the importance of nuclear weapons for France, Tertrais therefore posits that some sort of 'extraordinary' shift in the international political environment would be required to push France towards zero, so that disarmament is an 'extreme hypothesis'.[19] To imagine the emergence of an international political environment in which there is 'no foreseeable major threat' to French 'vital interests' is to imagine, according to him, a 'profound transformation of international relations' approaching some sort of 'global democratic peace' where the use of military force is constrained by international law.[20]

Tertrais also excludes the possibility of domestic political forces pushing for zero because, he argues, France 'has never had a significant anti-nuclear movement'. Furthermore, today the Green party 'is the only significant force calling for nuclear disarmament' and the French public is in favour of retaining a nuclear arsenal.[21] Having reviewed three possible scenarios by which France might 'reduce to zero', Tertrais concludes that:

> the only credible circumstances where France would be willing to seriously consider the global abolition of nuclear weapons are those in which there is no foreseeable major threat against its vital interests and those of its European partners. However, it would be difficult for Paris to stay away from a coordinated US-Russia-China initiative to begin negotiations for a treaty to eliminate nuclear weapons from all nations.[22]

To unpack this a little, absent domestic political pressure altering the power elite's cost/benefit analysis regarding the value of nuclear weapons, they will continue to be deployed to protect France's 'vital interests'. The needs of French national security are thus an unquestioned constant, so that nuclear deterrence is justifiable and pragmatic. Such *security first* logic, driven by a belief in French exceptionalism, provides a useful insight into how France's establishment—at the helm of a 'leading nation' with a 'civilizing mission'— sees itself.[23] It is therefore posited by Tertrais that France's nuclear needs will only diminish when other states act to 'roll back' proliferation and Russia becomes a democracy 'in the Western camp'.[24] The continuation of US extended deterrence to Europe is also seen as necessary to provide the requisite security guarantees. Alternatively, if the major nuclear powers delegitimise and relinquish their arsenals, this would lead to 'strong pressures from within the EU for France to follow suit'.[25] The belief that the US and Russia have prime responsibility for moving the world towards zero is a long-standing French position, alongside the need for these two nations to establish limits on defensive systems, for example, BMD and reductions to conventional forces.[26]

## 6.3 Critical perspectives on the causes and consequences of French nuclear possession and disarmament

This section follows Sagan in arguing that the realist understanding of French nuclear possession outlined above 'does not stand up very well against either existing evidence or logic'.[27] The reasons why the security model is insufficient shall be outlined below and *institutional democratisation* shall be explored as a means of providing an improved explanation of the causes and consequences of French nuclear possession and disarmament. Supporting evidence for *institutional democratisation*, which focuses on French domestic politics, is strong and can be found in several scholarly and expert works. For example, whilst contemporary political analysts may disagree over whether France should continue to possess nuclear weapons or commit to disarmament, there is agreement over the fact that French nuclear weapons decision-making has always been made by a tiny group of officials, initially technocrats and then political elites, for whom these weapons are of supreme importance.

Tertrais therefore observes that France may be 'the only country whose political system proceeds from the possession of nuclear weapons'.[28] Tertrais also follows Samy Cohen's description of France as a 'nuclear monarchy', and states that nuclear weapons policy making has actually become more centralised since the bomb was acquired.[29] Meanwhile, one French disarmament activist I interviewed stated that nuclear deterrence is 'in the DNA' of France's top political elites.[30] Elsewhere, Beatrice Heuser makes the complementary point that the bomb presents a unifying feature of national life in the French Fifth Republic as part of successive government's 'drive towards centralisation'.[31] Securing a national consensus on the 'two pillars of France's defence, conscription and the bomb', was therefore, Heuser argues, 'crucial for domestic reasons'.[32]

As previously noted, France's nuclear arsenal—which came to be known as the *force de frappe*, meaning strike force—is strongly associated with the legacy of Charles de Gaulle, who established the Fifth Republic and was its first President. Endowed with the authority to use nuclear weapons by the 1958 Constitution, De Gaulle dominated French foreign and defence decision-making for more than a decade.[33] Yet Heuser provocatively argues that, rather than De Gaulle's acquisition of the bomb enjoying popular support, during his presidency,

> the majority of the French population was against the development of a national nuclear force, and debates about nuclear weapons in the National Assembly and in the press were heated. De Gaulle carried out his costly programme against formidable opposition.[34]

Moreover, according to Gabrielle Hecht and Hugh Miall, the key decisions allowing the French to test a nuclear bomb in 1960 were taken by the Administrator General of the Commissariat à l'énergie atomique (CEA), Pierre Guilliaumat in the 1950s.[35] Guilliaumat pushed for the production of weapons grade plutonium, seeing the CEA's work as a symbol of 'technological prowess' and a means by which France could regain the 'national radiance' it had lost following the humiliation of WW2.[36]

Crucially, the high degree of financial and political autonomy enjoyed by the CEA during the unstable years of the Fourth Republic, which saw 20 prime ministers come and go, meant that the decision by the French government to acquire the bomb in 1960 only ratified a pre-existing bureaucratic process.[37] Where De Gaulle broke with the previous regime and left his own mark, according to Kohl, was in making France's nuclear arsenal principally a 'political instrument to support his independent foreign policies which sought to change the European and international system and France's role in it'.[38] To this extent, an approach to the French bomb informed by the principles of *institutional democratisation* is able to accept and incorporate the idea of De Gaulle seeking nuclear weapons for nationalistic reasons, but would also highlight the domestic political goals behind his and other elite's decision-making. Thus, rather than the bomb just enhancing France's political and military position within NATO, De Gaulle sought to use the new technology to restore French 'grandeur', both to prevent France moving from being a 'world empire' to 'a backward colonized nation' and to inspire and unite the French people in support of a revived state apparatus.[39]

However, the end of the Cold War represented a 'critical juncture' for French elites as one of the main pretexts for substantial military spending and nuclear deterrence—the Soviet threat to Europe—disappeared overnight.[40] The security model's explanation regarding why France continued to possess nuclear weapons thus faces a serious challenge here. Yet, as Heuser notes, France's nuclear deterrence policy has historically been aimed 'in all directions of the compass' and thus, on the international level, related to long-term power balancing rather than ephemeral ideology.[41] This includes 'resisting the domination of the US' which, she states, has been seen by 'governments since de Gaulle' as 'France's principal rival'.[42] Moreover, Paris only joined the NPT in 1992, having previously cited this treaty's 'hegemonic' nature.[43]

Yet France's desire for greater European integration from the 1980s onwards was increasingly at odds with its competitive 'nuclear nationalism' and its relationship with NATO needed to be resolved. Proponents of the nuclear force, such as Debouzy, therefore argued in the mid-1990s that, unless France rethought the purpose of its nuclear weapons—for example, by giving it a 'European role', including cooperation with the UK and Germany to develop a new system of deterrence—it might 'slowly fade into irrelevance'.[44] Whilst France and the UK have become increasingly close partners in this field, leading to a 2010 treaty on nuclear cooperation, European states have indicated their preference for the security guarantees offered by the US, under the auspices of NATO, to the idea of relying on France for a 'Europeanized' nuclear deterrent.[45] At the same time, EU member states have in recent years also developed a Common Foreign and Security Policy prioritising non-proliferation and disarmament.[46]

As noted above, during Chirac's Presidency nuclear weapons took on an enlarged role for France. Whilst in 1996 Chirac decided to dismantle the Pierrelatte uranium enrichment plant and fissile material production facilities, as well as sites for ground-based missiles, he also restarted nuclear testing in 1995, in the face of much domestic and international criticism, including opposition from 60% of the French public, before deciding to ratify the CTBT in 1998.[47] Indeed, the French government's approach to nuclear tests reveals much about the attitude of it and other NWS to the CTBT and the idea of nuclear disarmament more generally. For example, in 1994, French Prime Minister Balladur said that the CTBT 'must not in any way envisage the elimination of nuclear weapons or seek to undermine the status of the nuclear powers'.[48]

In terms of recent nuclear politics and the prospects for disarmament, Presidents Sarkozy and Hollande reiterated the need for nuclear weapons, the former describing them as the 'nation's life insurance policy' and the latter as both a 'protection against all threats' and 'an element that fosters peace'.[49] Former Assistant Chief of the UK's Defence Staff, Rear Admiral John Gower, has observed that through the modernisation of its nuclear arsenal, including 'a new class of SSBN, the replacement of nuclear-capable combat aircraft, and the upgrade of both its submarine and air-launched nuclear-armed missile capabilities', France has signalled its intention to 'retain a nuclear capability well into the 21st century'.[50] Despite this, several scholars have argued that the future of France's nuclear status is in jeopardy. For example, Tertrais recently stated that, whilst France will remain strongly committed to the bomb for the foreseeable future, this direction will become more difficult because 'the ability to maintain and adapt the French deterrent is weakening'.[51] He therefore concludes that 'serious political will, as well as significant resources human, technological, budgetary' will be necessary if France is to retain its nuclear arsenal.[52]

Regarding the domestic social and political forces within France that are supportive of nuclear disarmament, the lack of democracy, transparency and accountability surrounding the French nuclear weapons system from its birth has significantly hampered public awareness and engagement. For example, former French defence minister Paul Quiles has described how neither the French parliament nor the mainstream media have sought to engage in a serious debate concerning nuclear weapons. He thus proposes that a 'fake consensus' exists regarding the *force de frappe,* given that the issue is handled with 'silence, approximations, counter-truths, slogans, authoritarian arguments', amounting to a 'French fib'.[53] This assessment, whilst covering one policy area, nonetheless sits uncomfortably alongside evaluations provided by the Economist Intelligence Unit and Freedom House, which find that France generally scores highly across a range of indicators concerning democratic standards and civic rights.[54]

Heuser helps us understand why the French 'fib' exists, highlighting the French ruling elite's 'fear' of 'strong popular disapproval' and 'internal discord' given the nation's 'stormy political history', which may lead to a divergence between the public and the President's 'will'.[55] In order to explore whether such a divergence exists today, the next section discusses whether the French citizenry are more apathetic or disapproving concerning the bomb and what this means for the French political system—in particular the Presidency. For example, whilst the majority of the French political establishment remains firmly wedded to nuclear weapons, there are some signs—such as recent support for nuclear disarmament from retired political and military officials, civil society and public opinion—indicating that anti-nuclear voices may be growing louder in France.

## 6.4 French nuclear politics today

Having introduced the perspective of *institutional democratisation* in relation to the history of the French bomb, we may now outline a fuller analysis to explain how this concept applies to modern-day French nuclear politics. For example, drawing on relevant scholarship and expert commentary, we will review the national and international political developments that could enable France to move towards nuclear disarmament over the medium to long term. Before discussing these processes, it is crucial to emphasise again that the French political system is set up so that nuclear weapons decision-making is highly centralised within the office of the President. Thus, when anyone describes 'Paris' or 'France' making a decision on nuclear weapons, they are really talking about a 'power elite'—comprised of 'a handful of political leaders and officials', as Tertrais notes.[56]

To consider how political processes—domestic and international—might lead to French nuclear disarmament is therefore also to consider both how political pressure and persuasion might alter the cost/benefit analysis of the small group that controls French nuclear weapons decision-making, and how the French political system might itself change as a result of the disarmament process. For example, the elimination of the French nuclear arsenal would clearly require Presidential acquiescence or active support given that any President who made such a move would be divesting their office of immense physical and symbolic power built up over several decades, implementing a far-reaching change to the political structure of the Fifth Republic and diverging from Washington and London's long-held positions. Domestic opposition to any disarmament initiatives would also likely emanate from powerful institutions with a stake in the nuclear weapons business, namely the 'nuclear community', including the CEA, the 'defence community', including the Ministry of Defence and armed forces and 'delivery systems manufacturers'.[57]

Turning to the prospect of changes to France's domestic political scene, one French disarmament activist I spoke to proposed that France could only achieve the complete elimination of its nuclear arsenal 'as a result of parliamentary pressure and public opinion'.[58] Yet, as previously mentioned, proponents of disarmament such as Quiles and civil society groups such as Mouvement De La Paix have drawn attention to the 'fake consensus' on nuclear issues in France, which has been brought about by a system in which there is little opportunity for societal debate or parliamentary influence over nuclear weapons decisions.[59] As Tertrais himself acknowledges, material support from Washington for the French nuclear programme was kept quiet in order not to interfere with the Gaullist narrative of nuclear weapons being a symbol of French independence, a myth widely recognised as crucial to the reconstruction of France's identity as a global power after WW2.[60]

In terms of national self-image today, most of the political establishment takes pride in France being a leading military power, remaining one of the world's largest spenders on defence as well as a major arms exporter.[61] As its 2017 Defence and National Security Strategic Review indicates, France sees itself as having responsibilities on several fronts, which require strong conventional military capabilities, in addition to a nuclear arsenal.[62] France has therefore sought to take a lead on the EU's Common Security and Defence

Policy, take an increased responsibility for security in Africa and contribute to the stability of the Middle East and Persian Gulf.[63] According to opinion polls, recent French military interventions in these areas have received growing and majority public support, as has the idea that France continues to act forcefully on the world stage, which may be explained by deep anxiety regarding Islamist terrorism and the perceived effectiveness and legitimacy of these actions.[64]

However, scholars also argue that, for French defence and foreign policy makers, 'the end of the Cold War symbolized the ultimate failure of the Gaullist and nationalist nation state identity' and that 'the chapter of Gaullism in French history is now closing'.[65] These elites have instead adopted a European identity and closer integration with EU partners, a move that has facilitated a return to NATO's military command structure—with public support, and cooperation with the UK on nuclear weapons.[66] More recently, as Rebecca Johnson notes, President Macron called for a discussion in Europe concerning how France's nuclear arsenal can 'provide the nuclear deterrence role in an integrated EU defence policy'.[67] This move was driven by Paris eyeing an opportunity in the post-Brexit political environment to achieve its long-held ambition of supplanting NATO nuclear weapons with European capabilities.

Yet, as described above, France has also sought to retain a strong conventional military profile, leading to discontent in the military (particularly the Army) and in Parliament concerning what Tertrais describes as the 'heavy burden of nuclear expenses in the defence budget', which were heightened by the austerity measures taken after the 2008 economic crisis.[68] These post-Cold War trends could help explain why one recent opinion poll indicated that the French public may be more opposed to nuclear possession than is commonly thought.[69]

In addition, as Heuser points out, opinion polls since 1980 have shown that large sections of the French public have been 'opposed to the use of nuclear weapons in defence of France, even if foreign forces were invading French soil'.[70] Other forms of opposition to nuclear weapons came in the form of the 2009 statement in favour of a NWFW by former Prime Ministers Alain Juppe and Rocard, Former Defence Minister Alain Richard, and retired General Bernard Norlain.[71] However, some, such as Thérèse Delpech and Venance Journé, have argued that these (and earlier) interventions have not made a significant impact on the public or the powers that be.[72] This may be because, as former French foreign minister Hubert Vedrine notes, the French public 'do not follow foreign policy very closely' and those issues which do resonate 'come down to a few images and symbols'.[73]

Other groups that have become more active on nuclear issues include Parliamentarians for Nuclear Non-Proliferation and Disarmament (PNND). For example, in May 2014 two PNND representatives briefed France's National Assembly Defence Committee on the 'economy and utility of nuclear weapons, the need to re-evaluate nuclear deterrence, and the universal obligation to achieve nuclear disarmament', the first such briefing by members of civil society since the establishment of the Fifth Republic.[74] Whilst this initiative is a small step it does correspond with the argument that domestic political pressure towards nuclear disarmament will likely need to be part of a larger democratising force, given the lack of sustained public attention given to this issue. For example, this could be part of the more general decentralisation of decision-making on defence and foreign policy, so that they are not solely based in the Élysée Palace.

Such a wider democratic revival is also shown to be necessary judging by recent studies concerning the health of French democracy. For example, in 2015 the Economist Intelligence Unit downgraded France in its Democracy Index to the status of a 'flawed democracy'.[75] Problems with the French political system are also shown in recent polling figures from Ipsos which found that 'only 8% of voters have confidence in political parties, fewer than one in five trust MPs and only 28% the institution of parliament', whilst 'nearly eight in 10 agreed that the system of democracy malfunctions in France as it isn't representative of voters' ideas'.[76] Given such popular dissatisfaction with France's current political model and ambivalence regarding nuclear possession and use, the seeds of a social movement to emerge capable of implementing *institutional democratisation*—which may directly or indirectly support nuclear disarmament—exists, but will need to be carefully cultivated over the medium to long term if the many obstacles built into the French social and political establishment are to be successfully challenged and overcome.

## 6.5 Conclusion

This chapter has explored different views on French nuclear politics in order to ascertain the strengths and weaknesses of the mainstream, realist and *institutional democratisation* perspectives on the causes and consequences of French nuclear disarmament. Ultimately, for this case study, the security model has very limited value in explaining French decision-making in this area because the Soviet nuclear threat was by no means the only or most important factor motivating France's pursuit of the bomb. Moreover, the end of the Cold War and improvements in relations between the West and Russia opened up the question of why France, if the security model was accurate, should keep hold of its nuclear arsenal. In response, supporters of French deterrence argued that France and its vital interests continued to face serious and diverse threats from abroad. Some also argued that the end of the Cold War and the move to European integration altered France's political and security environment in a fundamental way, so that the original Gaullist design for the bomb as a revisionist tool for changing the European and international system, and France's independent role in it, had been superseded.

French nuclear deterrence is thus now, according to its proponents, intended to contribute to the security of NATO and Europe through war prevention, whilst budgetary pressures and the desire for a fellow NWS in Europe have also pushed France into nuclear cooperation with the UK. Yet despite their case for continued nuclear possession appearing to be significantly undermined following the end of the Cold War, advocates of the bomb—such as Tertrais—appear not to be concerned about the near-term prospects of French nuclear disarmament. This is at least in part because they believe that the elimination of the French bomb would require far-reaching and far-off changes in international politics and security and because of the limited political support for disarmament that exists within France.

Whilst authors such as Tertrais take us further than most other mainstream and realist authors in examining how domestic and international politics interact to drive French nuclear possession and stymie nuclear disarmament efforts, the analysis provided by *institutional democratisation* remains valuable in helping us imagine the circumstances by which France may eventually relinquish the bomb. To begin with, focusing on the domestic front allows us to consider the political aims and interests of French decision-making elites in relation to their acquisition of the bomb. For example, after the national trauma of WW2, the leaders of the Fifth Republic sought to re-establish France's position as an independent, sovereign nation—strategically separate from the US—and unite the nation behind the Presidency. Nuclear possession was seen as a means to accomplish all these goals—despite significant public and party-political opposition to the bomb. Whilst some mainstream and realist works consider these ambitions, they are not presented as primary factors for France possessing the bomb, but rather as secondary issues.

*Institutional democratisation* does a better job here by emphasising the integral role that nuclear possession plays in the French political system—particularly the office of the President—as summed up by the widely accepted concept of France being a 'nuclear monarchy'. Given the structural implications of the bomb for French governance, it is clear that the incrementalism proposed by the *guardianship* approach to disarmament is insufficient, so that there will likely need to be significant political change on the domestic and international fronts if the French nuclear arsenal is to be eliminated. The concept of *institutional democratisation,* as applied to this case study, also focuses on the ideas and beliefs of France's power elite, the majority of whom believe that nuclear weapons continue to be the cornerstone of national security, helping to ensure continuity and stability at home—and thus prevent a return to political turmoil.

Financial and political resources have therefore been made available to modernise and optimise the *force de frappe* to ensure its deployment for the foreseeable future. The continuity of French nuclear weapons policy also reflects the more general continuity of French national strategic culture as proposed by Heuser. French decision-makers see France as a global power with attendant responsibilities based on their nation's 'vital interests', which require the ability to project power to key areas of the world. But they also recognise the important role this grandiose self-image can play in binding the nation together behind the President, given his or her role as the 'nuclear monarch'.

Warnings that France's nuclear arsenal may fall victim to structural disarmament or simply fade into irrelevance, would therefore appear to have been heeded and addressed by those interested in and responsible for its reproduction, though it is unclear for how long the French nuclear weapons complex can continue to be supported, given its size and expense. In addition, there are some signs of a growing questioning of France's nuclear status amongst elites and the public. Meanwhile, the pragmatic changes to optimise France's nuclear force, partly chosen and partly forced upon decision-makers by circumstance, have been sold to the world, rather disingenuously, as examples of France's commitment to disarmament.

Given France's political system and its situation internationally, sincere French progress towards complete nuclear disarmament in the long term would therefore likely require some combination of the following: substantially increased domestic agitation by civil society, focused on improving the political system's democracy, transparency and accountability in order to initiate public and parliamentary engagement on nuclear issues; increased political pressure for France to commit to nuclear disarmament from its key European partners and global civil society; further reductions to the nuclear arsenals of Russia and the US.

## Notes

1 Sagan, Scott D. (1996) Why Do States Build Nuclear Weapons? Three Models in Search of a Bomb, *International Security,* 21(3), Winter, 58, 76–77.
2 Heuser, Beatrice (1998) *Nuclear Mentalities? Strategies and Beliefs in Britain, France and the FRG* (London: Palgrave Macmillan), 79.
3 Sagan, Why Do States Build Nuclear Weapons?, 77.
4 Kohl, Wilfrid L. (1971) *French Nuclear Diplomacy* (Princeton, NJ: Princeton University Press), 8.
5 Journé, Venance (2011) France's Nuclear Stance: Independence, Unilateralism, and Adaptation, in McArdle Kelleher, Catherine and Reppy, Judith, eds., *Getting to Zero: The Path to Nuclear Disarmament* (Stanford, CA: Stanford University Press), 126; Tertrais, Bruno (2004) Destruction Assurée: The Origins and Development of French Nuclear Strategy, 1945–81, in Sokolski, Henry D., ed., *Getting MAD: Nuclear Mutual Assured Destruction: Its Origins and Practice* (Carlisle: SSI), 61; Yost, David (2004) France's Nuclear Deterrence Strategy: Concepts and Operational Implementation, in Sokolski, *Getting MAD*, 224.
6 Kohl, *French Nuclear Diplomacy*, 6; Tertrais, Destruction Assurée, 58.
7 Debouzy, Oliver (1995) A European Vocation for the French Nuclear Deterrent, in *United States Air Force, Western European Nuclear Forces: A British, a French and an American View* (Santa Monica, CA: RAND), 34; Harrison, Michael M. (1981) Consensus, Confusion and Confrontation in France: The Left in Search of a Defense Policy, in Andrews, William G. and Hoffman, Stanley, eds., *The Impact of the Fifth Republic on France* (Albany, NY: SUNY), 261.

8  Heuser, *Nuclear Mentalities*, 81.
9  Yost, France's Nuclear Deterrence Strategy, 209.
10  Chirac, Jacques (2001) Speech at the Institut des Hautes Études de Défense Nationale, www.elysee.fr, 8 June; Yost, France's Nuclear Deterrence Strategy, 218; Journé, France's Nuclear Stance, 141.
11  Acronym Institute (2006) Chirac Reasserts French Nuclear Weapons Policy, *Disarmament Diplomacy*, 82, Spring.
12  French Republic (2013) *Defence White Paper*, www.defense.gouv.fr, 73.
13  Yost, France's Nuclear Deterrence Strategy, 223.
14  Ibid., 219.
15  Acronym Institute, Chirac Reasserts.
16  Ministère de la Défense (2009) *Nuclear Disarmament: France's Concrete Commitment*, onu.delegfrance.org.
17  Ibid.
18  French Republic (2017) *Defence and National Security Strategic Review*, www.defense.gouv.fr, 15.
19  Tertrais, Bruno (2009) French Perspectives on Nuclear Weapons and Nuclear Disarmament, in Blechman, Barry, ed., *France and the United Kingdom* (Washington, DC: Henry L. Stimson Center), 16–23.
20  Tertrais, Bruno (2009) *The Political and Strategic Conditions of Nuclear Disarmament*, www.frstrategie.org, 4
21  Tertrais, Bruno (2007) The Last to Disarm: The Future of France's Nuclear Weapons, *Nonproliferation Review*, 14(2), July, 261.
22  Tertrais, French Perspectives, 19.
23  Moran, Matthew and Cottee, Matthew (2011) Bound by History? Exploring Challenges to French Nuclear Disarmament, *Defense and Security Analysis*, 27(4), December, 344.
24  Tertrais, Last to Disarm, 269; Tertrais, French Perspectives, 17.
25  Tertrais, French Perspectives, 18.
26  Yost, David S. (1994) France, in Murray, Douglas J. and Viotti, Paul R., eds., *The Defense Policies of Nations: A Comparative Study*, 3rd Edition (Baltimore, MD: Johns Hopkins University Press), 266.
27  Sagan, Why Do States Build Nuclear Weapons?, 77.
28  Tertrais, Last to Disarm, 257.
29  Cohen, Samy (1986) *La monarchie nucléaire: Les coulisses de la politique étrangère sous la Ve République* (Paris: Hachette).
30  Interviewee: H.
31  Heuser, *Nuclear Mentalities*, 90.
32  Ibid., 91.
33  Tertrais, Last to Disarm, 257.
34  Heuser, *Nuclear Mentalities*, 98.
35  Hecht, Gabrielle (1998) *The Radiance of France: Nuclear Power and National Identity After World War II* (Cambridge, MA: MIT Press), 63; Miall, Hugh (1987) *Nuclear Weapons: Who's in Charge?* (Basingstoke: Macmillan), 64.
36  Hecht, *Radiance of France*, 63.
37  Ibid., 74; Hymans, Jacques (2012) *Achieving Nuclear Ambitions: Scientists, Politicians, and Proliferation* (Cambridge: Cambridge University Press), 34.
38  Kohl, *French Nuclear Diplomacy*, 6.
39  Hecht, *Radiance of France*, 62.

40 Marcussen, Martin, Risse, Thomas, Engelmann-Martin, Daniela, Knopf, Hans Joachim and Roscher, Klaus (1999) Constructing Europe? The Evolution of French, British and German Nation State Identities, *Journal of European Public Policy*, 6(4), 614–633.

41 Heuser, *Nuclear Mentalities*, 127, 143.

42 Ibid.

43 Becker, Una, Müller, Harald and Wisotzki, Simone (2008) Democracy and Nuclear Arms Control—Destiny or Ambiguity? *Security Studies*, 17(4), 826.

44 Debouzy, European Vocation, 37, 69.

45 Sloan, Stanley R. (1997) *French Defense Policy: Gaullism Meets the Post-Cold War World*, www.armscontrol.org, April.

46 Grand, Camille (2010) France and Disarmament from one Century to Another, www.diploweb.com, 25 July.

47 Moran, Matthew and Cottee, Matthew (2011) Bound by History? Exploring Challenges to French Nuclear Disarmament, *Defense and Security Analysis*, 27(4), December, 347.

48 Jabko, Nicolas and Weber, Steven (2007) A Certain Idea of Nuclear Weapons: France's Nuclear Nonproliferation Policy in Theoretical Perspective, *Security Studies*, 8(1), 145.

49 *New York Times* (2008) Sarkozy Announces Small Cut in France's Nuclear Arsenal, www.nytimes.com, 21 March; Hollande, Francois (2012) François Hollande: La dissuasion nucléaire est 'un élément qui contribue à la paix', www.lemonde.fr, 21 June; Nuclear Threat Initiative (2013) France Won't Give up Nukes: President, www.nti.org, 9 January.

50 Gower, John (2018) *Nuclear Signalling between NATO and Russia*, www.europeanleadershipnetwork.org, 6.

51 Tertrais, Last to Disarm, 270.

52 Ibid.

53 Quiles, Paul (2010) *Nuclear Weapons, a French Fib: Reflections on Nuclear Disarmament* (Paris: Éditions Charles Léopold Mayer), 20.

54 Economist Intelligence Unit (2016) *Democracy Index 2016: Revenge of the 'Deplorables'*, ww.eiu.com, 7; Freedom House (2017) *Populists and Autocrats: The Dual Threat to Global Democracy*, https://freedomhouse.org, 21.

55 Heuser, *Nuclear Mentalities*, 90–91.

56 Tertrais, Last to Disarm, 258; Nectoux, Francois (1986) France, in McLean, Scilla, ed., *How Nuclear Weapons Decisions Are Made* (Basingstoke: Macmillan), 184.

57 Nectoux, France, 154.

58 Interviewee: H.

59 Mouvement de la Paix (2014) Qui Sommes-Nous?, www.mvtpaix.org/wordpress; Quiles, *French Fib*, 5.

60 Tertrais, Bruno (2011) US-French Nuclear Cooperation: Stretching the Limits of National Strategic Paradigms, www.wmdjunction.com, 26 July.

61 SIPRI (2020) *Arms Transfers Database*, www.sipri.org.

62 French Republic (2017) *Defence and National Security Strategic Review*, www.defense.gouv.fr, 56.

63 Watanabe, Lisa (2013) France's New Strategy: The 2013 White Paper, *CSS Analysis in Security Policy*, 139, September, 1.

64 Chevalier, Guillaume (2011) From Afghanistan to Syria, Current State of Public Opinion as Regards Defence Issues, www.ifop.com, 18 June; Dinmore, Guy (2011)

Poll Shows Little Support for Libya Intervention, www.ft.com, 4 May; de Durand, Etienne and Pertusot, Vivien (2013) Defense Matters: IFRI Contribution on France, www.ifri.org; Dahlgreen, Will (2015) 3 in 4 French Support Intensifying Air Strikes on ISIS, https://yougov.co.uk, 22 November.

65  Marcussen, Martin, Risse, Thomas, Engelmann-Martin, Daniela, Knopf, Hans Joachim and Roscher, Klaus (1999) Constructing Europe? The Evolution of French, British and German Nation State Identities, *Journal of European Public Policy*, 6(4), 630; Zaretsky, Robert (2010) End of the French Exception, www.mondediplo.com, November.

66  Crumley, Bruce (2009) Sarkozy Moves to Restore France's NATO Role, http://content.time.com, 11 March; Moran, Matthew and Cottee, Matthew (2011) Bound by History? Exploring Challenges to French Nuclear Disarmament, *Defense and Security Analysis*, 27(4), December, 341–357.

67  Johnson, Rebecca (2020) Macron's Post-Brexit Nuclear Ambitions Are Destined to Fail, www.theguardian.com, 10 February.

68  MacLachlan, Ann and Hibbs, Mark (2006) Chirac Shifts French Doctrine for Use of Nuclear Weapons, www.globalresearch.ca, 12 February; Tertrais, Bruno (2006) *Memorandum from Dr Bruno Tertrais*, www.publications.parliament.uk, 17 February; Collin, Jean-Marie (2013) The French White Paper on Defence and National Security, www.basicint.org, 14 May.

69  IFOP (2012) Presidential Election: The French and Military Spending, www.ifop.com

70  Heuser, *Nuclear Mentalities*, 88.

71  Juppe, Alain, Norlain, Bernard, Richard, Alain and Rocard, Michel (2009) Pour un désarmement nucléaire mondial, seule réponse à la prolifération anarchique, MM. Juppé, Norlain, Richard et Rocard, www.lemonde.fr, 14 October.

72  Delpech, Thérèse (2005) French Nuclear Policy: More Continuity than Change, in Huntley, Wade L., Mizumoto, Kazumi and Kurosawa, Mitsuru, eds., *Nuclear Disarmament in the 21st Century* (Hiroshima: Hiroshima Peace Institute); Journé, France's Nuclear Stance, 140.

73  Ford, Peter (2005) On World Stage, France's Role Is Audience Favorite, www.csmonitor.com, 30 September.

74  PNND (2014) PNND France Coordinator Briefs Parliament Defense Committee, www.pnnd.org.

75  Economist Intelligence Unit, *Democracy Index 2016*, 46.

76  Nardelli, Alberto (2015) From Margins to Mainstream: The Rapid Shift in French Public Opinion, www.theguardian.com, 8 January.

# 7 China

## 7.1 Introduction

This chapter assesses the explanatory power of *institutional democratisation*, alongside mainstream and realist approaches, regarding the causes and consequences of Chinese nuclear possession and disarmament. As with the British and French case studies, the chapter proceeds over several sections, beginning by comparing and contrasting different views on China's nuclear experience since the 1960s, before moving on to discuss modern-day Chinese nuclear politics. Whilst there is a growing range of data available in English covering Chinese nuclear politics, the size and scope of these sources remains modest in comparison to the other NWS case studies, for several reasons. These include the language barrier, the limited size of the readership for this subject outside a specialist Chinese and Western audience, and the understandably closed nature of this subject given China's authoritarian government. In addition, the Chinese nuclear arsenal is much smaller than that of Russia and the US, and China's power projection capabilities and ambitions are also relatively limited. Overall, China therefore plays a growing but still lesser role in most other state's strategic calculus, particularly those outside its immediate region. For all these reasons, this chapter is considerably shorter than those on the US and Russia in this study.

Section 7.2 examines the mainstream and realist approaches explored in Chapter 2 to assess the merits of the security model in relation to China's particular experience as an NWS. This is principally done by placing China's development of nuclear weapons within the context of relevant historical events from the mid to late 20th and early 21st century—focusing, in particular, on the Cold War, where China positioned itself as a regional power capable of deterring threats to its independence and security, including via its nuclear status. It is important to note upfront that the evidence for *institutional democratisation* is less compelling for the Chinese example compared to the other NWS case studies. This is first because the security model, as utilised by mainstream and realist works, explains how the perceived US threat drove the Chinese decision to acquire the bomb and continues to be the primary driver of China's deterrence strategy. Moreover, unlike Russia and the US, China has maintained a relatively small nuclear arsenal for several reasons. These include Beijing's hitherto limited international ambitions and resources and its belief in minimal deterrence as a sufficient defensive strategy. This meant that the bomb did not take on the same kind of significance for Chinese elites and domestic politics as the other NWS, something which can also partly be explained by Beijing's gradually less hostile attitude to the US and retreat from a revolutionary international policy over the course of the Cold War.[1]

Section 7.3 then assesses the explanatory power of *institutional democratisation* in relation to the causes and consequences of Chinese nuclear possession and disarmament. In doing so, official justifications for China's nuclear status are explored and the limitations in the mainstream and realist literature's explanation of Chinese nuclear politics considered. To the extent that *institutional democratisation* and the domestic politics model are usefully applicable, I draw on the work of experts in Chinese nuclear history and politics to specify the wider impact nuclear possession has had on modern China. More specifically, I introduce discussion of how domestic elite actors and groups shape Chinese defence and foreign policy, including on nuclear issues, to examine the barriers to institutional change pursuant to disarmament. Having discussed different theories of Chinese nuclear possession and disarmament, section 7.4 reviews modern-day Chinese politics in relation to nuclear matters and goes into more detail concerning the current obstacles to and opportunities for *institutional democratisation* in China. Chapter 8 then contextualises China's responsibility for nuclear disarmament action alongside the four other NWS.

## 7.2 Mainstream and realist perspectives on the causes and consequences of Chinese nuclear possession and disarmament

Mainstream and realist explanations of the origins of Chinese nuclear possession principally focus on external security challenges. Analysts such as Lu Hui and Scott Sagan explain that China began its nuclear programme in response to nuclear threats from the US during the Korean War and the later Taiwan Straits crisis in the mid-1950s.[2] Elsewhere, Nicola Horsburgh states that these factors led to 'Chinese attitudes towards nuclear weapons' changing 'dramatically in the early to mid-1950s' as Beijing's 'technological weakness' was exposed by Washington's 'development of tactical nuclear weapons, the thermonuclear hydrogen weapon and the fusion-fission weapon'.[3] Moreover, around this time, the US chose to deploy its tactical nuclear weapons to bases near China, 'in South Korea, Taiwan, Guam, and Hawaii'.[4] As Avery Goldstein notes, deteriorating Sino-Soviet relations during the 1960s further increased the perceived value of a nuclear arsenal for Beijing because of the 'limited value of China's conventional deterrent'.[5]

Marshal Nie Rongzhen, the head of China's science and technology complex from 1958 to 1967, stated in his memoirs that China chose to develop nuclear forces over conventional weaponry in order to put an end to China's 'period of being bullied, humiliated and oppressed'.[6] Importantly, Chinese strategists did not intend to use nuclear weapons to intimidate or coerce others. This was because, according to the strategic thought of leaders such as Mao Zedong and Deng Xiaoping, nuclear weapons were 'paper tigers' which could not achieve specific military objectives during wartime, so that, as Mao once said, 'with only atomic bombs and without people's struggles, then atomic bombs are meaningless'.[7]

Chinese strategists continue to view the threat of US military power—conventional and nuclear—as the main reason to possess nuclear weapons. Stability and continuity in the face of such external threats are thus among the key defining features of China's approach to the bomb. This approach has also been informed by the Communist Party leadership's strategic plans and insights over several decades. As noted above, China's aim in becoming a nuclear power was to break the US and Soviet Union's 'great power monopoly on nuclear weapons' and avoid coercion.[8] Indeed, retired Major General Pan Zhenqiang states that China has always calibrated its nuclear posture in response to 'the threat posed to it by the United States' nuclear strategy'.[9]

The key strategic factors for China today concern the US's military presence in East Asia and its policy of containment. Given the US's conventional superiority, China thus fields its nuclear arsenal as part of a central deterrence strategy. Yet China's recent moves to secure and seize territory in the South China Sea and the long-running dispute over Taiwan's status have also taken on a nuclear aspect. Analysts such as Jing-Dong Yuan have therefore noted the need for greater dialogue between Washington and Beijing, given the significant 'misperceptions and misunderstandings' between the two nations concerning nuclear weapons, deterrence and strategic stability, and how these issues might play out with regard to regional disputes.[10]

Some, such as James Acton, point out that today's China looks north to Russia and south to India regarding nuclear and other dangers.[11] Others argue that China does not see Russia as a threat, but rather as benign, and a partner it can work with to counter the US, though Beijing is concerned about possible US–Russia cooperation on BMD because of Russia's ability to monitor Chinese ballistic missile launches.[12] In addition, the re-emergence of Japan as a regional power, which is remilitarising, has the potential to build a nuclear weapon, is part of the US's extended deterrence network and which cooperates with the US on BMD, is of particular concern to Chinese planners.[13] Yet the fact remains that it is principally the US's threatening behaviour and its array of technologically advanced military capabilities that make Chinese decision-makers consider qualitative and quantitative improvements to their nuclear forces. The key question here is China's need to ensure a secure second-strike capability, which leads it to prioritise ambiguity, secrecy and mobility regarding its nuclear weapons system.

Moreover, as Jeffrey Lewis notes, China's 'force deployments and arms control behavior' both suggest that the Chinese leadership is convinced that 'even a very small, unsophisticated force maintained a measure of deterrence against larger, more sophisticated nuclear forces'.[14] Thus, since acquiring the bomb in 1964, China has continued to possess only a small number of nuclear weapons, which have been 'based largely on a single mode of delivery, kept off alert and under the most restrictive declaratory posture—a categorical no-first-use pledge'.[15] China has invested significant sums in developing advanced military technology, although, as Lewis explains, it did not produce as many nuclear bombs as its 'resources, material, manpower and industrial capacity' allowed, because its leaders believed that a larger arsenal would not enhance deterrence.[16]

Today, China's estimated stockpile of 290 nuclear warheads is thus the fourth smallest of the NWS.[17] China's 2019 Defence White Paper describes how it is 'always committed to a nuclear policy of no first use' whilst pursuing 'a nuclear strategy of self-defense' and 'does not engage in any nuclear arms race with any other country'. The paper also states that China 'advocates the ultimate complete prohibition and thorough destruction of nuclear weapons'.[18] China's 2010 Defence White Paper included proposals for how nuclear disarmament may be advanced, with Russia and the US identified as bearing 'special and primary responsibility' for this task given the size of their arsenals. China's participation in 'multilateral negotiations on nuclear disarmament' will then occur 'when conditions are appropriate'.[19] Chinese analyst Wu Zhan has suggested that such conditions would include Russian and US nuclear arsenals being reduced by 90–95%, plus an end to testing and production, though this action may only persuade China to agree to keep its arsenal at low numbers.[20]

## 7.3  Critical perspectives on the causes and consequences of Chinese nuclear possession and disarmament

As discussed above, the evidence for the security model concerning the causes and consequences of Chinese nuclear possession (and, to an extent, disarmament) make it, for this study, the outlier in terms of the relative explanatory power of *institutional democratisation*. At the same time, however, there are limits to the mainstream and realist explanations discussed in Chapter 2 that need to be recognised and engaged with. For example, the previous section showed that China—which has historically seen itself as the superior 'Middle Kingdom' and regional hegemon—had its status severely downgraded over several centuries. The threat posed by the US following the Chinese Communist Revolution in 1949 should therefore be seen in relation to China's historic sense of external danger. Yet the benefits of nuclear possession, for Chinese decision-making elites, can also partly be explained by domestic politics. For example, by acquiring the bomb, China's Communist leadership recovered a sense of national pride and regional prowess, demonstrating the new regime's leadership abilities to the populace.

To develop our evidence base concerning *institutional democratisation* and its relevance to the Chinese case, we therefore need to draw on the work of subject specialists who highlight the political importance of nuclear weapons for Beijing on the international *and* domestic levels. For example, according to Lewis and Horsburgh, proponents of the Chinese nuclear programme in the 1950s and 1960s believed that the pursuit of the bomb would have benefits beyond security.[21] In addition to uniting the country behind Beijing's rule, the bomb would help China become an advanced nation by developing its economy, science and technology.

For Horsburgh, Beijing therefore saw possession of the bomb as 'crucial' in 'improving China's legitimacy and prestige domestically, regionally, and internationally'. Such 'imperatives' can, for this author, be seen both in China's early nuclear efforts and the more recent modernisation of its military.[22] It is also possible to consider the international relevance of *institutional democratisation* to China given Beijing's foreign policy. For example, Shaun Breslin argues that 'in China's view, there is still a lot to be done before the institutions of global governance become truly representative and democratic'. Until international institutions become more equitable, China will thus likely see its nuclear arsenal as an important way of retaining a seat at the top table of global affairs.[23]

As noted in the previous section, China's nuclear arsenal has been subject to a high degree of opacity which, the Chinese government argues, is vital to maintain its strategy of keeping this arsenal at low numbers. Thus, whilst nuclear secrecy is common across NWS, in addition to China's status as a one-party state without formally democratic institutions, its particular nuclear strategy further heightens the need to keep outsiders in the dark, which includes ensuring its citizens remain uninformed on this subject.

The lack of a free press or political opposition in China means that the public has no opportunity to play a role in nuclear weapons debates. One 2009 poll conducted by the *People's Daily* newspaper states that '51% of respondents wanted nuclear disarmament while 49% did not'.[24] Yet it is reasonable to question how representative and useful such data are, given how poor the condition of civil liberties is in China. National decision-making on security in China remains highly centralised in the top leadership as a matter of course. Yet, according to one Chinese nuclear expert I spoke to, at higher levels of civil society, for example, in academia and specialist media, there does exist some critical discussion of nuclear issues and future strategy which can have an influence on the thinking of ministers.[25]

Whilst Chinese nuclear policy making is thus particularly opaque and difficult to map, it is possible to outline the key Chinese institutions with a stake in the nation's nuclear weapons system. These consist of the Chinese Communist Party (CCP), the leadership of the People's Liberation Army (PLA) and the defence-industrial and scientific community.[26] Although political power in China has generally become 'diffuse, complex, and at times highly competitive', according to Kerry Dumbaugh and Michael Martin, nuclear weapons policy has remained closed and rigid.[27] For example, Bates Gill and Evan Medeiros argue that Chinese nuclear weapons decision-making has consistently been dominated by 'one person or of a small clique of key political individuals', including key members of the Central Military Commission of the CCP, which oversees the PLA.[28]

These authors also suggest that the influence of the PLA over 'the formulation and operationalization of nuclear doctrine' and the 'R&D and procurement process for nuclear weapon, missile and command-and-control systems' has increased in recent times.[29] This influence is also likely to grow as China's nuclear forces expand in 'size, technical sophistication' and 'mobility', which has implications for nuclear weapons decision-making as 'tensions could arise' between civilian and military constituencies, for example, over doctrine and the size of China's nuclear arsenal. Overall, Gill and Medeiros conclude that Chinese decision-making on nuclear weapons is 'best understood as being under civilian control but lacking democratic accountability'.[30]

Increased military influence over the future direction of China's nuclear arsenal may well make arms control negotiations in this field more challenging, given that, as the authors of a report by the James Martin Center note, the Chinese military is 'more suspicious of nuclear disarmament concepts'.[31] This is despite the fact that nuclear weapons have significant opportunity costs for China, given its high levels of poverty and relatively low spending on human development priorities such as health and education.[32] In terms of the prospects for *institutional democratisation* for the Chinese case, we may therefore conclude that, in the short to medium term, it is first necessary that China has a civilian rather than a military *guardianship* in control of nuclear weapons policy if future disarmament action is to be possible.

## 7.4  Chinese nuclear politics today

After the review of different theories concerning how and why China acquired the bomb, as well as the barriers to Chinese nuclear disarmament, this section shall consider the state of Chinese nuclear politics today. In doing so we shall further assess the utility of *institutional democratisation* for providing insight and understanding into how China might commit to nuclear abolition. As discussed in the previous sections, whilst there are several countries of importance to the future evolution of China's nuclear arsenal, the key determining factor is the behaviour of the US and its allies in East Asia. In addition to external influences, there are also important domestic factors—such as bureaucratic and political interests—driving China's nuclear weapons decision-making.

In recent years there have been several in-depth studies, of both Western and Chinese origin—some of which have already been mentioned—which shed light on how the interaction between these internal and external factors have created the Chinese nuclear weapons system as it is today. These studies provide important insights into the domestic and international political obstacles that need to be overcome, and the opportunities that need to be seized upon, if China is to move towards nuclear disarmament. Studies of particular note, informing the discussion below, include: the James Martin Center for Nonproliferation Studies report *Engaging China and Russia on Nuclear Disarmament,* which includes contributions from several authors, Lora Saalman's *China and the U.S. Nuclear Posture Review* and retired Chinese Major General Pan Zhenqiang's analysis *China's Nuclear Strategy in a Changing World Strategic Situation.* In addition, Shen Dingli, Bates Gill and Evan Medeiros, and Hui Zhang have made notable contributions to this debate.

A useful point of departure for our discussion is Pan Zhenqiang's observation that China 'is not a nuclear weapon state in the traditional Western sense'. This is because China does not see nuclear weapons as 'essential instruments to help achieve political aims', whereas the US and Soviet Union used them to 'intimidate other countries or control their allies', including, for Washington, through extended deterrence.[33] As previously discussed, the UK and France also see nuclear possession as a means of maintaining some semblance of a global strategic outlook and influence with Washington, with both arsenals linked to power projection and overseas intervention. Whereas the four other NWS have, to different degrees, seen nuclear weapons as a means of realising their global political ambitions, China—as a much weaker power—has had domestic economic development as its main priority since the early 1980s.[34] China's nuclear strategy and policy has therefore, hitherto, been configured in line with these means and ends, so that unless and until its strategic position significantly changed, nuclear weapons would continue to exist for central deterrence, requiring only a small, survivable force.

Whilst China's recent nuclear modernisation, which has brought it near to achieving the ability to deliver nuclear strikes from land, sea and air, may therefore appear alarming, as Joshua Pollack notes, the 'planned enhancements to the nuclear force' included in China's 2015 white paper on military strategy, reflects Beijing's focus on 'incremental changes to operational practices' rather than an intent to engage in an arms race with Washington.[35] Such modifications to China's nuclear posture are, Pollack argues, driven by US advanced conventional capabilities, which, 'together with US ballistic missile defense programs, Chinese experts regard as a threat to China's ability to conduct nuclear retaliation'.[36]

Liping Xia therefore argues that, because China needs 'a long-term peaceful international environment, especially stable surroundings', in order to meet its national development objectives, it will not act in ways that 'seriously disturb the current international economic and political mechanisms except when its critical national interests are threatened'.[37] Moreover, according to the US DOD, China's need for international stability means that it will 'avoid direct confrontation with the United States and other countries'.[38] If China were to act aggressively this would jeopardise its push to modernise because it requires 'extensive economic and technological cooperation with the outside world, both with its neighbors and with the West', according to Tiejun Zhang.[39] Yet China is also currently, for Rex Li, attempting to escape Washington's grip so that it can 'eventually become an economic superpower and a global strategic player' capable of replacing 'US domination' with 'multipolarity'.[40] In practice, as Noam Chomsky argues, Beijing is thus looking for ways to escape the US's 'arc of containment in the Pacific', which limits China's 'control over the waters essential to its commerce and open access to the Pacific'.[41]

As evidence of Beijing's growing assertiveness, analysts point to President Xi Jinping's 2017 speech to the nation's Congress, in which he declared that China was ready to 'take centre stage in the world'.[42] Such pronouncements, for Christopher Layne, signal that Beijing's 'days of biding its time and hiding its capabilities are over'.[43] Moreover, in 2018 China's legislature removed presidential term limits, opening the path for Xi to remain as China's ruler indefinitely. In terms of China's development of advanced military technology, its focus on A2/AD is based on, Luis Simon argues, its particular 'geographical setting', whereby it seeks to 'block access to its long and open littoral'.[44] Beijing has therefore developed a range of satellites and a growing selection of short- and medium-range cruise and ballistic missiles, allowing it to hit strategically important targets across East Asia. As discussed in Chapter 3, China has also become a leading developer of hypersonic weapons, with the delivery of nuclear warheads in mind, has the ability to damage key US capabilities with its anti-satellite missiles and is a leading cyber power. Overall, Beijing's growing technological sophistication means that it is increasingly able to challenge the US militarily, and is starting, as Layne notes, to 'draw level with the United States in *regional* military power in east Asia'.[45]

The implications of China's transition towards great-power status is far-reaching for the US, global and regional order. Layne thus argues that, since China is not 'converging with the West: it is not going to become a democracy any time soon—if ever', and, unless the US can adapt to China's new position, the prospects of war between the two nations 'are high'.[46] As previously discussed, if current tensions are to dissipate—pursuant to nuclear arms control and disarmament action—the US therefore needs to recognise China's interests in East Asia. This means that Washington's strategy of extended deterrence will need to be rethought, alongside the nature of its security relationships with regional allies such as Japan and South Korea. Moreover, key disputes, for example, regarding Taiwan, the Korean peninsula and territorial and maritime demarcation, will need to be resolved as part of reaching sustainable security agreements.[47]

Without such agreements, it is highly unlikely that China will move towards eliminating its nuclear weapons. This is because, as Saalman highlights, Chinese analysts see 'self-determination' and 'the belief that disarmament must not threaten a country's independence, sovereignty, or security' as amongst the core principles guiding China's approach to arms control and disarmament.[48] Given that Beijing considers Taiwan to be part of China, nuclear disarmament is seen by Beijing as compromising China's vital interests. This is because the fate of Taiwan is intertwined with both the legitimacy of the CCP and US regional military commitments and is thus, according to analysts from the Center for Strategic and International Studies, the area where nuclear weapons 'would most likely become a major factor'.[49]

Additionally, on the domestic front, if China is to transition to a non-nuclear identity as a FNWS, its leaders will need to make strides in delegitimising nuclear weapons as a source of national self-esteem. This is because nuclear weapons compensate somewhat for its strategic military imbalance with Western powers by providing a sense of pride in the nation's technological prowess.[50] According to one Washington-based analyst I spoke to, China also takes pride in having overcome various difficulties in order to achieve its nuclear status. Yet this has purposefully been a quiet type of pride, which has not translated into a deeper nuclear culture—potentially making the process of delegitimising nuclear weapons easier.[51] Moreover, the long-standing diplomatic support China has given to nuclear disarmament and a global ban on nuclear weapons may ease an eventual transition to FNWS status.[52]

In terms of domestic political changes that may support *institutional democratisation* and smooth the path towards disarmament, Minxin Pei argues that, whilst the CCP learned lessons from the fall of the Soviet Union and the Tiananmen Square protests, China's authoritarian one-party state, like all others, has a limited shelf life. Moreover, China's recent economic and social development puts it 'well into' a 'zone of democratic transition', so that democracy, he argues, could arrive through several different routes.[53] Indeed, for Vladimir Frolov, China is presently 'more democratic than Russia', so that whilst the latter is 'faking democracy to cover up an emerging dictatorship' the former is 'evolving into a more pluralistic system'.[54]

However, for Wei-Wei Zhang, it is 'unimaginable' that the Chinese people 'would ever accept' a 'multi-party democratic system'.[55] Thus, as with the Russian populace, democracy in China will require the Chinese people to develop into an active and engaged citizenry with an enlarged social consciousness. Clearly, any such internal transformation will benefit from a stable regional security environment, both so that civilians rather than the armed forces can exert control over key policy areas and to enable military restraint—including on nuclear issues.

## 7.5 Conclusion

This chapter has sought to assess the extent to which *institutional democratisation,* in comparison with other established theories, provides insight and understanding into the causes and consequences of Chinese nuclear possession and disarmament. Ultimately, for the Chinese case study, the security model is largely persuasive when outlining the origins of China's nuclear weapons programme. For example, China acquired its nuclear weapons to both counter the threats and coercion it faced from the US since the 1950s and put an end to China's long history of humiliation and oppression from outside forces. China's nuclear weapons system was then shaped by factors including the top leadership's assessment of trends in international security, resource constraints and the need for domestic economic development.

The utility of the domestic politics model can be seen when China's national pride in possessing the bomb is taken into account. Over time, bureaucratic and institutional interests found it useful to maintain the nation's nuclear status because the CCP gained legitimacy from having re-established China as a regional power. The limited nature of China's strategic ambitions during the Cold War and focus on internal development is also reflected in the restrained character of its approach to nuclear weapons, including its minimum deterrence policy.

Mainstream and realist approaches are also useful in identifying the international security requirements for nuclear disarmament. Despite the political, scientific and technological benefits of attaining nuclear status for the CCP, Chinese nuclear possession does not seem to play as important a role in the nation's political system, economy, society and elite identities when compared to NWS with similarly sized nuclear arsenals, such as France and the UK. Thus, as General Pan Zhenqiang pointedly observes, because China has avoided using the bomb to further its international political aims, it 'is not a nuclear weapon state in the traditional Western sense'.[56]

At the same time, because China—like Russia—is a regional hegemon seeking independence from Western containment, it does highly value its nuclear arsenal, albeit in a different way to Moscow. This is, again, primarily because of the historic restraint shown by China, which led to it developing a relatively modest nuclear force focused on central deterrence. In terms of disarmament, China's restraint and disarmament rhetoric may thus facilitate an easier path to eventual abolition. However, given that China is an authoritarian one-party state, it is also clear that there are far greater barriers to *institutional democratisation* than for the Western democracies that are nuclear possessors, so that whilst it could play an important role in realising Chinese nuclear disarmament, democratisation may only develop over the medium to long term.

In order for China to eventually move towards eliminating its nuclear weapons, it is first necessary for it to continue with its policy of restraint and to resist further building up its arsenal. The potentially escalating confrontation with the US and other powers in the Asia-Pacific region endangers this and requires attention at the highest levels, with the two nations using sustained diplomacy to achieve a political settlement. Furthermore, Chinese analysts have presented a range of concerns and proposals that need to be addressed—principally by the US, but also the other NWS and Japan—if Beijing is to cooperate on any process for nuclear reductions or disarmament. Absent international cooperation focused on making progress in these areas there will be little incentive for China's political leadership to alter its long-standing approach to nuclear policy, especially given the current absence of significant domestic pressure which could make the CCP reassess their options and change course.

# Notes

1  Horsburgh, Nicola (2015) *China and Global Nuclear Order: From Estrangement to Active Engagement* (Oxford: Oxford University Press), 158.
2  Wortzel, Larry M. (2007) *China's Nuclear Forces: Operations, Training, Doctrine, Command, Control and Campaign Planning* (Carlisle: Army War College), 28; Sagan, Scott D. (1996) Why Do States Build Nuclear Weapons? Three Models in Search of a Bomb, *International Security,* 21(3), Winter, 58–59.
3  Horsburgh, *China and Global Nuclear Order,* 41.
4  Ibid., 42.
5  Goldstein, Avery (1992) Robust and Affordable Security: Some Lessons from the Second-Ranking Powers during the Cold War, *Journal of Strategic Studies,* 15(4), 494.
6  Lewis, Jeffrey (2004) The Minimum Means of Reprisal: China's Search for Security in the Nuclear Age, doctoral dissertation, https://drum.lib.umd.edu, 238.
7  Fravel, M. Taylor and Medeiros, Evan S. (2010) China's Search for Assured Retaliation: The Evolution of Chinese Nuclear Strategy and Force Structure, *International Security,* 35(2), 62.
8  Ibid., 71.
9  Zhenqiang, Major General Pan (2009) China's Nuclear Strategy in a Changing World Strategic Situation, in Blechman, Barry, ed., *China and India* (Washington, DC: Henry L. Stimson Center), 33.
10  Yuan, Jing-Dong (2009) China and the Nuclear-Free World, in Hansell, Christina and Potter, William C. eds., *Engaging China and Russia on Nuclear Disarmament* (Monterey, CA: James Martin Center for Nonproliferation Studies), 35.
11  Acton, James M. (2012) Bombs Away, Being Realistic about Deep Nuclear Reductions, *Washington Quarterly,* 35(2), 40.
12  Sokov, Nikolai N., Yuan, Jing-dong, Potter, William C. and Hansell, Cristina (2009) Chinese and Russian Perspectives on Achieving Nuclear Zero, in Hansell and Potter, *Engaging China and Russia,* 16; Interviewee: E.
13  Kristensen, Hans M., Norris, Robert S. and McKinzie, Matthew G. (2006) *Chinese Nuclear Forces and U.S. Nuclear War Planning* (Washington, DC: FAS/NRDC); Hughes, Christopher W. (2009) *Japan's Remilitarisation* (London: IISS); Saalman, Lora (2011) *China and the US Nuclear Posture Review* (Washington, DC: Carnegie Endowment for International Peace), 35.
14  Lewis, Minimum Means, 12–13.
15  Lewis, Jeffrey (2009) Chinese Nuclear Posture and Force Modernization, in Hansell and Potter, *Engaging China and Russia,* 38.
16  Lewis, *Minimum Means,* 239–240.
17  Kristensen, Hans M. and Korda, Matt (2019) Chinese Nuclear Forces: 2019, *Bulletin of the Atomic Scientists,* 75(4), 172.
18  People's Republic of China (2019) *Defence White Paper,* 9.
19  People's Republic of China (2010) *Defence White Paper* (Beijng: PRC), 74.
20  Saalman, Lora (2009) How Chinese Analysts View Arms Control, Disarmament, and Nuclear Deterrence After the Cold War, in Hansell and Potter, *Engaging China and Russia,* 69.
21  Lewis, Minimum Means, 239–240.
22  Horsburgh, *China and Global Nuclear Order,* 75.
23  Breslin, Shaun (2013) China and the Global Order: Signalling Threat or Friendship? *International Affairs,* 89(3), 631.

24  Qiang, Guo (2009) US Nuke-Free World Plan Stirs Debate, *Global Times*, 24 September.

25  Zhang, Hui (2012) China, in Acheson, Ray, ed., *Assuring Destruction Forever: Nuclear Weapon Modernization around the World* (New York: Reaching Critical Will/WILPF), 25; Interviewees: I, J.

26  Gill, Bates and Medeiros, Evan S. (2010) China, in Born, Hans, Gill, Bates and Hanggi, Heiner, eds., *Governing the Bomb: Civilian Control and Democratic Accountability of Nuclear Weapons* (Oxford: Oxford University Press), 130.

27  Dumbaugh, Kerry and Martin, Michael F. (2009) *Understanding China's Political System* (Washington, DC: Congressional Research Service), 2.

28  Gill and Medeiros, China, 130.

29  Ibid., 151.

30  Ibid., 150.

31  Sokov et al., Chinese and Russian Perspectives, 7.

32  Page, John and Thakur, Ramesh (2013) *Nuclear Weapons: The Opportunity Costs* (Canberra: APLN/CNND), 7–8.

33  Zhenqiang, China's Nuclear Strategy, 30.

34  Shen, Dingli (2008) China's Nuclear Perspective: Deterrence Reduction, Nuclear Non-Proliferation, and Disarmament, *Strategic Analysis*, 32(4), 642.

35  China Power Project (2020) How Is China Modernizing its Nuclear Forces? https://chinapower.csis.org; Pollack, Joshua H. (2015) Boost-Glide Weapons and US-China Strategic Stability, *Nonproliferation Review*, 22(2), 160.

36  Ibid., 155.

37  Xia, Liping (2009) China's Nuclear Policy and International Order in the 21st Century, in *Major Powers' Nuclear Policies and International Order in the Twenty-First Century* (Tokyo: National Institute for Defense Studies), 87.

38  US Department of Defense (2014) *Quadrennial Defense Review*, www.defense.gov.

39  Zhang, Tiejun (2002) Chinese Strategic Culture: Traditional and Present Features, *Comparative Strategy*, 21(2), 84.

40  Li, Rex (2003), Changing China–Taiwan Relations and Asia-Pacific Regionalism: Economic Co-operation and Security Challenge, in Dent, C., ed., *Asia-Pacific Economic and Security Co-operation New Regional Agendas* (London: Palgrave Macmillan), 193.

41  McNeill, David (2014) Noam Chomsky: Truth to Power, www.japantimes.co.jp, 22 February.

42  BBC (2017) Xi Jinping: 'Time for China to Take Centre Stage', ww.bbc.co.uk, 18 October.

43  Layne, Christopher (2018) The US–Chinese Power Shift and the End of the Pax Americana, *International Affairs*, 94(1), 106.

44  Simón, Luis (2016) The 'Third' US Offset Strategy and Europe's 'Anti-Access' Challenge, *Journal of Strategic Studies*, 39(3), 130.

45  Layne, US–Chinese Power Shift, 95.

46  Ibid., 108.

47  Center for Strategic and International Studies (2013) *Nuclear Weapons and U.S.-China Relations, A Way Forward* (Washington, DC: CSIS), 5.

48  Saalman, Chinese Analysts, 51.

49  Lieber, Keir A. and Press, Daryl G. (2009) The Nukes We Need: Preserving the American Deterrent, *Foreign Affairs*, November/December, 43; Center for Strategic and International Studies, *Nuclear Weapons*, 5.

50 McLean, Scilla ed. (1986) *How Nuclear Weapons Decisions Are Made* (Basingstoke: Macmillan), 188; Zhenqiang, China's Nuclear Strategy, 51.
51 Interviewee: K.
52 Ministry of Foreign Affairs of the People's Republic of China (2013) Statement by Chinese Delegation on Nuclear Disarmament at the Second Session of the Preparatory Committee for the 2015 Review Conference of the Parties to the NPT, fmprc.gov.cn, 23 April.
53 Pei, Minxin (2008) How China Is Ruled, *The American Interest*, March/April, 46–51; Pei, Minxin (2013) The Doomed Transitional Moment of 1989, in Stoner, Kathryn and McFaul, Michael, eds., *Transitions to Democracy: A Comparative Perspective* (Baltimore, MD: Johns Hopkins University Press), 378–399.
54 Frolov, Vladimir (2013) China More Democratic than Russia These Days, https://themoscowtimes.com, 23 June.
55 Zhang, Wei-Wei (2012) *The China Wave: Rise of a Civilizational State* (Hackensack, NJ: World Century), 60.
56 Zhenqiang, China's Nuclear Strategy, 30.

# 8 Realising NWS nuclear disarmament through institutional democratisation

## 8.1 Introduction

The preceding chapters of this study identified the need to improve upon mainstream and realist theories of nuclear possession and disarmament by introducing the concept of *institutional democratisation*. In order to assess the explanatory power of *institutional democratisation* in relation to existing theory, I discussed these approaches alongside the five NWS case studies, exploring the relevant historical record and scholarship. Overall, it was found that the security model's emphasis on external threats to sovereignty and national security being the primary factor driving NWS's acquisition and continued possession of nuclear weapons has limited value for the cases of the US, UK and France, being more relevant to Russia and, in particular, China. Moreover, mainstream and realist theories do not account for the important domestic political factors driving the acquisition, reproduction and development of the bomb in each NWS over time. Crucially, mainstream and realist theories are also unable to provide a compelling account of how nuclear disarmament may be achieved and sustained, including what political forces will be required to realise NWS's transition to FNWS status.

In contrast, *institutional democratisation* is able to explain key aspects of nuclear possession and disarmament for the US, Russia, the UK and France, with China as a lesser included case. This is primarily because this approach provides insight and understanding into the domestic drivers of nuclear acquisition, reproduction and development. Furthermore, *institutional democratisation*—as a theory of political change—improves upon mainstream and realist theories by identifying the key political obstacles to nuclear disarmament at the domestic level and how these may be overcome through democratic reforms. It is also important to repeat that *institutional democratisation* does not reject the security model's insights into the international drivers of nuclear possession, but seeks to add to and improve upon them where appropriate.

Whilst the case studies covered in Chapters 3 to 7 discussed these different theoretical approaches to understand nuclear politics in and between NWS, it is also necessary to provide an overview of how the NWS relate to one another as a group and place the NWS within the context of the global nuclear order.

Doing so helps us to understand both how this group and this order operates and how they might, together, be transformed in ways supportive of nuclear disarmament. This analysis is presented below in order to: i) further show how the domestic politics of and strategic decisions made by NWS affect each other's nuclear choices, ii) produce ideas and strategies concerning how the elimination of nuclear arsenals might be advanced, focusing on the concept of *institutional democratisation* as it applies to both the national and international levels.

The chapter begins (in section 8.2) with a review of possible actions and processes supportive of *institutional democratisation* in the NWS—involving the state, civil society and the public, for example—and how these may be developed to realise the proposed disarmament measures. Section 8.3 considers where the political will for these changes might come from and what tools and strategies disarmament advocates may find useful to adopt. This involves a discussion of key related issues across NWS with implications for disarmament, including: recent trends in domestic and international politics, including public opinion, populism and the prospects for a TPNW; financial, safety and technological factors involved in nuclear modernisation; and the need to differentiate between states in terms of their responsibility for national nuclear disarmament and a NWFW. Section 8.4 highlights the importance of considering how other emerging global challenges may interact with the prospects for nuclear disarmament in future, with a particular focus on the politics of climate change.

## 8.2 Institutionalising civilian and democratic control over nuclear weapons systems at a national level

Our previous discussion has shown that nuclear disarmament at the national level, as with the transition to denuclearisation at the regional and international levels, will look different and feel different for each of the NWS given the singular nature, including the nuclear and political histories, traditions and institutions, of each of these states. For example, the far greater scale of the Russian and US nuclear arsenals and the supporting political, military and industrial infrastructure, means that the task of moving to FNWS status will be a much bigger societal undertaking for these two states than for China, France and the UK. On the other hand, as Alger and Findlay point out, France, Russia, the UK and US have experience of, and previously paid for, developing the facilities to dismantle the first generation of their nuclear weapon systems that they can draw on in future.[1]

Therefore, at the same time as being aware of the differences between NWS, it is also important to recognise the commonalities between these states, both because, unlike the vast majority of NNWS, their polities have been shaped in significant ways by the possession of nuclear weapons, and because of the types of activities involved in the process of eliminating nuclear arsenals. As discussed above, nuclear weapons systems are able to exist and be reproduced because political actors in each NWS have, over several decades, developed highly secretive and autocratic institutions that protect these weapons from popular political control. The citizens of NWS thus share the challenge of how they can develop governance processes, appropriate to their particular circumstances, that will allow nuclear weapons systems to be controlled, and their salience reduced, pending their elimination. This challenge may be eased, for example, through NWS (and other state and non-state actors) exchanging and sharing knowledge and ideas on disarmament practices, whilst safeguarding against proliferation by restricting access to sensitive information.

It should be noted, however, that specifically democratic domestic governance processes may be less important for Chinese nuclear disarmament, with the first priority being civilian rather than military control over nuclear weapons decision-making. Notwithstanding the particular nature of the Chinese case, it remains the fact that each NWS must move towards a setup whereby the domestic political conditions that allow nuclear weapons to flourish are no longer present or have been dramatically reduced. In this sense, *institutional democratisation*, as a critical approach to power structures, can be seen as a means of addressing the domestic political problems, such as the centralisation of decision-making power in the hands of small elite groups, created by nuclear possession. It is therefore necessary and useful to give a sense of the specific practical ideas and processes supportive of NWS *institutional democratisation* that may be adopted by scholars, policy professionals and others working in this field.

### 8.2.1  *Measuring democracy and the benefits of nuclear disarmament*

As discussed in Chapters 3 to 7, mainstream analyses and indexes of democratic standards and civil liberties in NWS do not take into account the domestic political impact of nuclear possession. Yet, as we have seen, the acquisition of nuclear weapons established highly centralised nuclear weapons decision-making processes in each NWS as part of their military and political establishments, which then became embedded in each polity over time, impairing the knowledge of citizens and preventing their involvement in a key area of national policy. Whilst the implications of this institutional development differ for each NWS, for example, according to their type of political system, level of democratic participation and the size and scale of the state's nuclear weapons complex, the importance of this subject requires further investigation. This is necessary both so that the impact of nuclear possession on governance processes may be ascertained, and so that nuclear issues are reflected in methodologies assessing the health of a state's democracy.

Measuring what it means for a state to: i) possess nuclear weapons and ii) maintain a sizeable military establishment, in terms of a state's regime type and democratic performance, could potentially be done by introducing what Richard Merritt and Dina Zinnes refer to as 'multi-dimensional' indices to categorise regime type.[2] This model, based on the work of Ted Gurr, Keith Jaggers and Will Moore, utilises several indicators to rank countries on 'two 0 to 10 point scales, one for institutionalized democracy and another for institutionalized autocracy' so that 'a country may thus rank high on one and low on the other ... or rank the same on both'.[3]

It is beyond the scope of this study to apply this model to the NWS, which could involve weighting the political footprints of their nuclear and military establishments in order to rank each of their overall political systems, for example, in terms of levels of democracy and authoritarianism. However, given the challenges involved in defining what democracy means in practice and the failure of mainstream methodologies to consider the impact of nuclear possession and military establishments on domestic political practices, such a multi-dimensional approach would be worth developing in future studies.

Including such factors may also bring more scrutiny to bear on the often invisible and unknown areas of a nation's political life and show how actions supportive of nuclear disarmament and demilitarisation might strengthen democratic practices in nuclear possessors, creating the possibility of a virtuous cycle of democracy and disarmament. Moreover, improving the accuracy of measuring regime type has implications for political and international relations scholarship, given the prominence of ideas such as Democratic Peace Theory. As discussed in section 2.7, identifying the power and influence of military and nuclear establishments in the decision-making processes of the three NWS (France, the UK and US) that have formally democratic institutions requires us to categorise them differently from other democracies, for example, as more elitist and less popular. Applying such a categorisation for regime type may both help us explain why one 'democratic' state engages in conflict, or supports arms control and disarmament, more or less than another, and how such behaviour may be changed, including through domestic political activism and reform, so that democratic institutions and processes become more popular, less elitist and more sympathetic to disarmament concepts.

### 8.2.2 Implementing policies supportive of institutional democratisation in NWS

In the 1980s Hugh Miall proposed a series of short-term measures that governments could take in support of moves towards nuclear disarmament. As well as being compatible with *institutional democratisation*, several of these proposals remain relevant today, particularly given the fact that all of the NWS are in the process of modernising their nuclear arsenals. They include: i) greater democratic, including parliamentary, powers to control spending and procurement decisions on conventional and nuclear weapons, ii) passing laws to prevent representatives of nuclear weapons laboratories and arms manufacturers from participating in nuclear weapons decision-making, iii) breaking up bureaucratic power that could resist popular disarmament initiatives by, for example, rotating officials to other departments, iv) exerting greater political control over nuclear weapons laboratories to prevent research and development on new weapons, v) upgrading the status of arms control and disarmament within government, for example, foreign and other specialist ministries to assess the impact of new weapons on 'existing and potential arms control agreements'.[4]

In addition to such bureaucratic and parliamentary measures, the centralised war-making powers of the executive, which reach their zenith in the NWS's 'nuclear monarchies', also need to be restrained, alongside the salience of nuclear weapons being reduced in national security policies, for example, as proposed by Kennette Benedict in relation to the US.[5] Supportive developments in the domestic sphere will be enhanced by progress made at the regional and international level, for example, involving conflict resolution, conflict prevention and cooperation on arms control and disarmament, although, as I discuss further below, these levels have varying degrees of importance for each NWS's attitude towards disarmament given each nation's different strategic circumstances.

### 8.2.3 Improving public understanding of and participation in nuclear disarmament in NWS

As previously noted, the size of each NWS's disarmament task will be affected by the *degree* of irreversibility that is deemed necessary, including in the eyes of the international community, if an NWS is to become a FNWS.[6] The required level of disarmament will naturally affect how long it takes and how much it will cost to, for example, decommission relevant weapons and equipment, dispose of fissile material, convert or scrap supporting military and industrial facilities and monitor and verify (through unilateral, bilateral and multilateral measures) these processes. In addition, if a speedier disarmament process is required this will likely escalate costs. Notwithstanding the 'paucity of data' concerning the costs of eliminating nuclear weapons, largely due to the classified or inaccessible nature of much relevant information—including for previous dismantlement efforts involving nuclear weapons and facilities—Alger and Findlay make the important observation that such costs 'pale in comparison to the financial burden of deploying, maintaining and upgrading nuclear arsenals in perpetuity'.[7]

Improved understanding, and communication to key audiences (for example, communities involved in or living near to nuclear-related facilities and trade unions involved in relevant production), of the practicalities of what nuclear disarmament entails, would benefit from studies being undertaken on the options for eliminating nuclear arsenals, including on defence conversion/diversification and who should pay for the different parts of the process. For example, NWS will need to conduct studies costing the dismantlement of their own arsenals, while multilateral bodies—Alger and Findlay suggest the International Atomic Energy Agency or a future 'Nuclear Disarmament

Commission'—will have to provide resources for multilateral monitoring and verification.[8]

At present amongst NWS, civil society groups in the UK, including Scottish CND, Nuclear Education Trust and the Nuclear Information Service, and in the US, including Stephen Schwartz of the Brookings Institution, as well as Susan Willett for the UN's Disarmament Research Institute, have undertaken the most detailed studies of this type.[9] Such research will need to be more widely disseminated and expanded across all NWS, including by mainstream and alternative media outlets, in order to develop common, shared understandings of the meaning and implications of eliminating nuclear arsenals. This is particularly necessary in relation to local community groups and trade unions, both to alleviate their concerns regarding their jobs and skills, and to increase their support for and participation in developing alternatives to nuclear weapons-related work as part of a wider conversion of industry away from military production and towards civilian goods and services.

### 8.3 Where will the political will for nuclear disarmament action come from?

In terms of the existing and potential sources of cooperative and progressive NWS action on nuclear disarmament that might take forward the proposals outlined above, Chapters 3 to 7 of this study looked in detail at the key question of whether the requisite political energy might derive from small groups of political elites—as per the *guardianship* model—or from broader social movements—as suggested by *institutional democratisation*. Overall, the historical record proves former Swedish Prime Minister Olof Palme correct in his assessment that 'It is very unlikely that disarmament will ever take place if it must wait for the initiatives of governments and experts. It will only come about as the expression of the political will of people in many parts of the world.'[10] Palme's insight is corroborated by the preceding analysis of the politics of the NWS's nuclear weapons systems showing that NWS military and political elites and the institutions they inhabit and maintain have, in general, acted as the principal barrier to meaningful progress on nuclear disarmament.

For example, France and the UK are part of a 'P3' grouping, led by the US, which cooperates at a high level, including through working together on nuclear weapons development (albeit on a bilateral rather than trilateral basis) and policy. The leaderships of the P3 see nuclear weapons as symbolic of their nation's world power status and at the apex of their global power projection capabilities. Chinese and Russian elites also see nuclear weapons as necessary for access to the top table, but, given their relative conventional military weakness and much more modest ambitions compared to the US, their nuclear strategies principally focus on central deterrence and regional influence to ensure regime survival, national sovereignty and independence.

The key point here is that the wider character of Western NWS's defence and foreign policy, characterised by what Paul Rogers refers to as the 'control paradigm', needs to change if nuclear weapons are to be delegitimised and disarmament (and all this entails politically) embraced.[11] This in turn requires domestic political movements with positive non-nuclear visions for national security and the strategies and strength to implement them on a consensual basis over the long-term. Such popular movements are especially necessary in the US given the great size of its nuclear weapons complex, which will take many years to dismantle, and the need for China and Russia to be convinced that the US is committed to and will not deviate from demilitarisation and disarmament so that international order is reformed on a more equitable and cooperative basis.

Yet, as discussed in Chapters 3 to 7, significant political obstacles currently exist preventing a democratic transition supportive of nuclear disarmament in each NWS. For example, Tariq Ali has observed that in recent years France and the UK have been ruled by an 'extreme centre', made up of political parties with largely indistinguishable platforms that, in Peter Mair's words, preside over 'hollowed out' democracies.[12] In the US meanwhile, Max Blumenthal argues that the Republican party has been 'shattered' following its capture by the radical right, with the Democrats now forming the moderate right.[13] Western political classes are thus disconnected from the citizenry and becoming 'appendages of the state', whilst the citizenry have lost faith in and abstain from electoral politics.[14] There are clear echoes of the West's predicament in Russia's own 'managed democracy' and the role the ruling United Russia party plays as an outgrowth of the state, balancing the interests of oligarchs and rallying the population around the flag in times of crisis. Meanwhile, power in China's one-party state is becoming more centralised, as President Xi Jinping 'steps forward as the strongman who defends the PRC's Leninist form of bureaucratic state capitalism', in the words of Jonathan Fenby.[15]

In response to the perceived failure of these political systems, as well as a mounting distrust in elites, populist and nationalist movements of very different types have risen in recent years espousing, in some cases, putatively anti-establishment politics. This phenomenon has been seen in the US with Donald Trump's Republican Party, in the UK with the UK Independence Party, in France with the Rassemblement National and in Russia with the Progress Party. Such movements clearly need to be treated on a case by case basis as many do not promote progressive values and policies compatible with *institutional democratisation* and all it entails for disarmament. Indeed, some of these movements either directly propose or contain prominent elements enthusiastic about maintaining or increasing military spending based on a belief in maximising national power.

Given that nuclear disarmament requires democracy and demilitarisation—at home and abroad—to be prioritised, emergent progressive political movements will therefore first need to be persuaded or pressured to adopt and maintain such principles by activists inside and outside of these groups. In recent years, political candidates with the potential to advance democratic reforms, demilitarisation and disarmament, if chosen to lead their countries, have been seen in the UK with Jeremy Corbyn as leader of the Labour party, and in the US with Bernie Sanders's campaign to be the Democratic Presidential candidate. Whilst their election bids were unsuccessful, the wider social movements Corbyn and Sanders represent will need to be strengthened if progressive forces are to defeat calls for rearmament and overseas intervention emanating from pro-nationalist and militarist groups in these countries.

### 8.3.1 Disarmament by other means? Financial, safety and technological factors

In addition to the development of popular movements within NWS focused on and capable of achieving disarmament action in the near term, it is important to note the impact that financial, safety and technological factors have on nuclear weapons systems. For example, the cost and time overruns endemic to hi-tech military systems, the difficulty of maintaining core industrial skills related to nuclear weapons, the numerous safety and security incidents nuclear weapons are prone to and the possibility that submarines or even ballistic missiles may be made obsolete by future technological developments, could together mean that the costs and risks of nuclear weapons become so great that nuclear disarmament becomes the pragmatic choice for one or more set of NWS decision-makers. Notwithstanding the fact that NWS governments have and will take steps to mitigate these problems, their cumulative impact certainly makes the process of reproducing nuclear weapons systems much more challenging.

At present the bureaucratic and technological momentum driving the nuclear enterprise in NWS seems capable of responding to threats to its future, although one cannot be certain that this will always be so. In any case, these financial and technical problems are interesting to consider in terms of what the domestic causes of disarmament might be, and what political consequences may result. To give one example of this, if a future UK government framed its decision to disarm as a normative choice then this would very likely lead to problems for its relationship with France and the US, particularly if London then called on Paris and Washington to eliminate their own arsenals because they now saw these weapons as morally unacceptable. Leaked diplomatic cables published in 2010 illustrated this dynamic, with Paris in particular objecting to moves by then British Prime Minister Brown and US President Obama to raise the profile of multilateral nuclear disarmament.[16]

Contrast this awkward situation with one where London framed a unilateral disarmament decision as a pragmatic technocratic choice based on cost grounds, with the resources from nuclear weapons being diverted to conventional forces—a stance which the UK's NATO allies would likely be much more understanding of and comfortable with. A technocratic disarmament scenario, whilst still of great moment, would also, potentially, minimize the impact on an NWS's bureaucratic and political structures compared to an explicitly normative scenario, as decision-making elites would likely seek to maintain their institutional power and realise their core strategic goals using other means.

In any case, a technocratic disarmament scenario is, at present, only plausible for France and, possibly to a greater extent, the UK, given these nations' economic and political situations. For Russia and the US, whilst technocratic concerns are of high importance with regards to the future functioning of their nuclear arsenals, these weapons' political and strategic value outweighs such concerns for decision-making elites. As for China, a technocratic disarmament scenario is currently implausible, given that Beijing is increasing its nuclear weapons capabilities and technological competence, branching out into new platforms and delivery systems.

Thus, notwithstanding the current democratic malaise, the fact remains that the best-case scenario for disarmament advocates would be for sufficient political pressure to eventually develop so that each NWS takes unilateral steps supportive of disarmament that are not dependent on reciprocation from other NWS and that are in line with their NPT obligations. Yet any such unilateral moves are currently treated as politically unacceptable by Moscow and Washington, with the numbers game—covering warheads and delivery systems—having particular symbolic importance, whereby each wants to prevent quantitative inferiority for domestic political as much as international strategic reasons. Beijing, London and Paris meanwhile argue that they have exercised restraint and either reduced their nuclear forces after the Cold War or kept them at low numbers, so that they will need to see substantial movement from the big two before they commit to further reductions or, in the case of China, capping the size of their nuclear arsenal.

### 8.3.2 *Differentiating responsibility for nuclear disarmament*

As previously noted, in order to look beyond the numbers game and consider the politics of nuclear weapons in the round, it needs to be accepted that each NWS has a dual disarmament responsibility. This means that in addition to having an obligation to achieve national nuclear disarmament, each NWS also has an obligation to act in ways supportive of a NWFW, thereby encouraging other NWS's efforts to move to zero. A key issue here is how best to designate NWS in terms of their relative power and the structural nature of their relations. For example, mainstream analyses split the NWS into two main categories, with Russia and the US as the 'big' nuclear powers and China, France and the UK as the 'small' nuclear powers. Given the size of their arsenals, Russia and the US have thus for several decades borne a 'special responsibility' for creating a NWFW.[17]

Whilst Russia's strategic power dramatically declined following the end of the Cold War, along with its ability to bargain with the US, China's rising economic and military strength drew it into an unstable 'great power' dynamic of cooperation, competition and potential conflict. Today, as Acton argues, if Russia and the US's reductions are to continue beyond New START, they want to see China, France and the UK join them in a multilateral process before they reach the level of the big two.[18] Although China's nuclear arsenal remains small and has not developed the same level of technical sophistication as the US and Russia, the potential for it to rapidly improve—both quantitatively and qualitatively—concerns planners in Moscow and Washington.

For example, Acton points out that 'the downward trajectory of the American and Russian arsenals risks colliding with the upward trajectory in China, India, and Pakistan'. He therefore proposes that 'the future evolution of the world's nuclear arsenals will depend principally on the interactions of five states' which 'form two triangles', so that 'the first consists of the United States, Russia, and China; the second: China, India, and Pakistan'. The key point being made here is that the process of reducing the big arsenals of the US and Russia beyond a certain level is now connected to regional powers with smaller arsenals, thus making the whole process much more complicated given the number of actors involved. Moreover, Acton's analysis sees France and the UK playing a peripheral role as their nuclear weapons are seen by Russia as 'extensions of the US arsenal'.[19]

Acton, Chalmers and the Deep Cuts Commission have therefore each recently investigated the political and technical conditions that would allow the US and Russia to move to low numbers of nuclear weapons, so that NPS arsenals contract and converge at a minimisation point, from which the more difficult move to zero might be made.[20] Whilst there is much merit in their analyses, the key issue remains how the political will allowing more substantial moves to be implemented may be summoned. The main weakness of these studies is thus that they share a *guardianship* theory of political change that is embedded within the status quo whereby think-tanks, in close proximity to governments, make recommendations to bureaucratic and political elites. The aforementioned authors thus do not seek to criticise the undemocratic practices of Western NWS governments, discuss the negative impacts of US hegemony on French, German and British democracy, or imagine how Western NWS's political systems might be changed as in the past, for example, through domestic pressure and reforms. Such democratic developments are disallowed, presumably, for being idealistic or naïve. Yet, ultimately, it is improbable that significant improvements to major power relations and progress on nuclear disarmament can be made without the arrival of domestic political forces capable of exercising control over the defence and foreign policies of Western NWS so that they are placed on a responsible and sustainable path over the long term.

Returning to Palme's point concerning 'the political will of people' around the world leads us to observe that citizens of NWS have a responsibility to take political action that reduces the salience of nuclear weapons in their nation's security policies and that supports their national disarmament obligations. Just looking at the size of Russia and the US's nuclear arsenals may lead one to conclude that citizens of these nations have a particularly great responsibility here. Yet it is also clearly necessary to be realistic about the ability of citizens in different countries to take meaningful political action given the level of freedom, including basic civil liberties such as free speech, in their country. If this argument is accepted, given the oppressive nature of the Russian state and Russia's relative military weakness on the one hand, and the US's more open and liberal society and military superiority on the other, it follows that US citizens have the greatest responsibility to move their nation towards nuclear disarmament and actions supportive of a NWFW.

In addition to looking at the health of pro-disarmament groups on a domestic level it is also necessary to look at the international level. For example, a 2008 opinion poll, conducted in 21 states around the world, found that 'in 20 countries, large majorities—ranging from 62% to 93%—favoured an international agreement for the elimination of all nuclear weapons'.[21] Yet, as Wittner explains, despite the majorities in favour of an international ban on nuclear weapons, 'such a ban was "strongly favoured" by only 20% of those polled in Pakistan, 31% in India, 38% in Russia, 39% in the United States and 42% in Israel'. He therefore concludes that 'it appears that today's public opposition to nuclear weapons, although widespread, does not always run very deep.'[22] It is useful to apply Wittner's analysis to the current political situation, for example, given that in July 2017, 122 nations voted in favour of a TPNW at the United Nations. This is first because it is unclear to what extent advocacy efforts for a TPNW grew out of previously existing public and elite support for abolition and secondly, because, in the short term at least, the prospects of this treaty being implemented, following its entry into force, appear negligible given the determined opposition of NPS and their allies.[23] Advocates of nuclear disarmament, whether in national or international groups, therefore find themselves in a difficult strategic situation.

The main challenge such groups face is how they can persuade or pressure NWS governments to reduce the salience of nuclear weapons in their security policies and fulfil their international legal obligations to disarm if they are not able to mobilise or draw on active and compelling public support. Such a means of pressure and influence has, in the past, been of great importance—such as in the 1980s when fear of nuclear war was a key factor in the nuclear disarmament movement reaching the peak of its powers. Yet several of the activists, officials and scholars I spoke to for this study agreed that this sense of fear dissipated following the end of the Cold War, significantly reducing the salience of nuclear issues for citizens.[24] Today, most nuclear disarmament campaigners are still based in Western nations where, for reasons of history and the relative freedom of their societies, there is greater potential to reach and organise people around these issues. Yet such groups often have limited resources and matters pertaining to nuclear weapons generally tend not to be amongst citizens' political priorities.

Moreover, prominent recent initiatives, such as the humanitarian approach to nuclear weapons, which was adopted by the International Campaign to Abolish Nuclear Weapons to build international support for the TPNW, provide a universal moral and juridical response to the illegitimate power of nuclear arsenals.[25] However, these initiatives, whilst registering some successes, may be criticised on the grounds that their analysis tends not to look beyond the weapons themselves and provide a well-rounded political appreciation of the domestic causes and consequences of nuclear disarmament across NWS, concerning institutional power and democratisation, for example. The need to differentiate between NWS based on their relative strategic power and behaviour, and thus their responsibility for disarmament, is also often not considered—a gap in the disarmament debate and these groups' political strategies that should be addressed in future studies.

One means by which such studies could do this would be to focus on addressing the 'responsibility gap' in the NPT, which has been previously raised by several expert nuclear analysts. For example, Sagan's article 'Shared Responsibilities for Nuclear Disarmament' explored the need to rebalance the responsibilities NWS and NNWS have for abolition under the NPT, so that these states can 'better honor their Article VI commitments'.[26] Sagan's proposals include NWS and NNWS funding both improved verification and monitoring of disarmament and enhanced IAEA safeguards for nuclear power facilities, in addition to negotiating international control of the nuclear fuel cycle. In response to this article, Jayantha Dhanapala questioned whether the responsibilities Sagan outlined are shared or 'asymmetrical', given that NWS 'have more capabilities', before concluding that 'Just as the "the polluter pays" principle applies to climate change, the NWS have the main responsibility for achieving a nuclear-weapons-free world'.[27] Whilst we should recognise the technical merits of Sagan's article, Dhanapala's approach to responsibility highlights the importance of equity and justice underpinning new disarmament frameworks.

For example, building on Dhanapala's analysis, we may propose that, in order for NWS to realise their commitment under Article VI of the NPT regarding an end to the nuclear arms race and a disarmament treaty, a new process or framework is required. This new framework could, I propose, be based on the principle of common rights and differentiated responsibilities (CRDR), which is adapted here from the United Nations Framework Convention on Climate Change (UNFCCC).[28] The UNFCCC's two large categories of signatory are based on its established principles of: i) 'common but differentiated responsibilities', so CBDR rather than CRDR, and ii) 'respective capabilities'. This model informs the Kyoto Protocol, which is an international agreement committing its parties to internationally binding emission reduction targets.

Additionally, CBDR and similar approaches have, as the German Development Institute explains, 'been put into practice in a variety of international regimes and policy arenas, including the World Trade Organization, the Montreal Protocol—on protecting the ozone layer—and the burgeoning debate on universal Sustainable Development Goals'.[29] The idea of differentiated responsibilities was first established in the 1992 Rio Declaration, where developed countries acknowledged they bear particular responsibility 'in the international pursuit of sustainable development in view of the pressures their societies place on the global environment and of the technologies and financial resources they command'.[30] In addition to and complementing the idea of differentiated responsibilities is the idea of common or universal rights. As the Center for Economic and Social Rights notes, 'International human rights law implies duties on states to respect, protect and support the fulfilment of all human rights, including economic, social and cultural rights outside of the country's territory.'[31]

Developing a CRDR-based nuclear disarmament framework presents significant analytical challenges, including how to construct a methodology which takes into account the interaction between nuclear and conventional weapons and how different types of military capabilities, postures and behaviour may be weighted from a qualitative, strategic perspective. Looking further forward, if such a rigorous and coherent framework has been produced, if it is to be agreed and implemented in any meaningful way, then it will clearly need to have been accepted by states, including the major powers, following a cooperative political process, with supportive domestic developments, including sustained public engagement. Notwithstanding these longer term political challenges, the ability to differentiate between all states (including groups of states such as the NWS) in terms of their responsibility for disarmament, would potentially be a powerful tool for advocates to have in the present, to better visualise and communicate the obstacles to and opportunities for a NWFW, as well as how possessor states can reduce their nuclear arsenals to converge on zero.

## 8.4  Twin existential threats

As should be clear from the discussion conducted above, the politics of nuclear disarmament are, in and of themselves, complex and multi-faceted. Such complexity is only heightened when we consider how nuclear politics interact with other major civilisational challenges and how these may impact on the prospects for disarmament. For example, the *Bulletin of the Atomic Scientists* has described climate change and nuclear war as the 'twin existential threats' facing humanity.[32] Elsewhere, several authors, such as Martin Rees and Toby Ord, have asked whether human civilisation will survive the 21st century given both these and other threats, including engineered pandemics and artificial intelligence.[33] Yet the depth and immediacy of the unfolding environmental crisis mean that it is particularly important to consider the impact global warming will have on nuclear weapons and disarmament politics in the near future.

In recent years there have been mounting and repeated warnings about the severe threats posed by climate change. In 2017, thousands of scientists from across the world co-signed a journal article stating that, despite the widespread degradation and destruction of the environment, 'humanity is not taking the urgent steps needed to safeguard our imperilled biosphere'.[34] In 2018 the International Panel on Climate Change published a landmark report warning that there are only 12 years for global warming to be kept to a maximum of 1.5C.[35] Even half a degree increase over this figure will significantly worsen the risks of drought, floods, extreme heat and poverty for hundreds of millions of people. Furthermore, the UN have stated that there could be 200 million climate refugees by 2050, whilst for researchers at Stanford University, 'intensifying climate change' will 'drive up' the risks of armed conflict.[36]

Based on these findings, if the governments of the wealthiest and most powerful nations persist with business as usual, responding to climate disaster through increased militarisation and a 'garrison state' mentality, this could lead to perpetual conflict, authoritarianism and a much-worsened outlook for NWS nuclear disarmament. For example, the UK's MOD have produced an analysis of the 'future operating environment' out to 2035, which states that the impacts of climate change, such as 'resource scarcity' could 'increase competition and act as a catalyst for intra- and inter-state conflict' as states seek to 'protect lines of communication' and 'guarantee access to resources'.[37]

The US DOD, Michael Klare argues, is also deeply concerned about these issues, because 'military leaders believe climate change seriously threatens U.S. national security' and is 'stirring up chaos and conflict abroad, endangering coastal bases and stressing soldiers and equipment'.[38] Senior US military figures, such as former Brigadier General Wendell Christopher King, have thus argued that climate change 'poses an increasing threat to peace and security in the world' since it acts as 'threat multiplier' and could 'indirectly increase risks of violent conflict'.[39] Such tensions could escalate in unpredictable ways between states, including to nuclear war. For example, nuclear-armed India and Pakistan share key water resources which are endangered by global warming. In this instance, as Sunil Amrith argues, humanity must hope that 'the existential importance of water' will 'defuse conflict as much as competing attempts to control water will deepen it'.[40]

Concerning the question of who benefits from preventing the types of transformational change required by global warming, Naomi Klein highlights how 'hardcore climate deniers' were drawn to the issue by an understanding of the domestic political implications of climate science, namely government regulation, public investment and wealth redistribution. Such 'deniers' have thus chosen to protect their 'neoliberal' ideology over the science.[41] We may draw a parallel with nuclear disarmament here, given the implications, outlined above, which the elimination of nuclear weapons will have for possessor states. NWS nuclear disarmament would involve political transformation at the national, regional and international levels, with far-reaching implications for economic, social, political and security policy. The scale and meaning of these changes, which, I have argued, will need to be based around *institutional democratisation*, are—as with climate change—so at odds with current elite thinking in NWS that they are resisted at every turn by decision-makers.

Yet if the global economy and society were reshaped by the most powerful nations to effectively respond to the threat posed by climate change, this would likely have benefits for progress on nuclear disarmament given that any progress towards more sustainable and democratic polities will provide opportunities for advances in related issues. For example, a Green New Deal, as proposed by several civil society groups in the NWS and beyond, could include steps to demilitarise societies and transition towards civilian goods and services, with military industry—including that which is involved in and related to producing nuclear weapons—converted to produce green technologies.[42] Prior to the 2020 US Presidential Election, the argument for rethinking security along these lines gained ground owing to the Covid-19 pandemic, with several US- and UK-based analysts highlighting the importance of investing resources in public healthcare, rather than military hardware.[43] Opinion polls from 2020 also show that the majority of people in China, France, Russia, the UK and US support a 'green economic recovery from Covid-19'.[44] It is clear that such public enthusiasm for progressive political change will need to be harnessed by civil society movements and decision-makers if NWS governments are to take unilateral and multilateral action to meaningfully address the twin existential threats of climate change and nuclear war.

## Notes

1 Alger, Justin and Findlay, Trevor (2009) *The Costs of Nuclear Disarmament*, http:// carleton.ca, September, 11.
2 Merritt, Richard L. and Zinnes, Dina A. (1993) Democracies and War, in Inkeles, Alex, ed., *On Measuring Democracy: Its Consequences and Concomitants* (New Brunswick, NJ: Transaction), 207.
3 Gurr, Ted, Jaggers, Keith and Moore, Will H. (1990) The Transformation of the Western State: The Growth of Democracy, Autocracy, and State Power since 1800, *Studies in Comparative International Development*, 25(1), March, 73–108.
4 Miall, Hugh (1987) *Nuclear Weapons: Who's In Charge?* (Basingstoke: Macmillan), 157–158.
5 Benedict, Kennette (2016) Add Democracy to Nuclear Policy, in Collins, Tom Z. and Wilson, Geoff, eds., *10 Big Nuclear Ideas for the Next President* (San Francisco, CA: Ploughshares Fund).
6 Cliff, David et al. (2011) *Irreversibility in Nuclear Disarmament: Practical Steps Against Nuclear Rearmament* (London: VERTIC), 27.
7 Alger and Findlay, *Costs*, 1.
8 Ibid., 2.

9   Ainslie, John (2012) United Kingdom, in Acheson, Ray, ed., *Assuring Destruction Forever: Nuclear Weapon Modernization around the World*, (New York: Reaching Critical Will/WILPF); Nuclear Education Trust (2012) *Trident Alternatives Review and the Future of Barrow*, www.nucleareducationtrust.org; Burt, Peter (2016) *AWE— Britain's Nuclear Weapons Factory: Past Present and Possibilities for the Future*, www.nuclearinfo.org; Schwarz, Stephen (2008) *The Costs of U.S. Nuclear Weapons*, www.nti.org, 1 October; Willett, Susan (2003) *Costs of Disarmament–Disarming the Costs: Nuclear Arms Control and Nuclear Rearmament* (Geneva: UNIDIR).

10  ElBaradei, Mohamed (2008) *Reviewing Nuclear Disarmament*, www.iaea.org, 26 February.

11  Rogers, Paul (2008) *Global Security and the War on Terror: Elite Power and the Illusion of Control* (Abingdon: Routledge), 207.

12  Ali, Tariq (2015) *The Extreme Centre: A Warning* (London: Verso); Mair, Peter (2006) Ruling the Void? The Hollowing of Western Democracy, https:// newleftreview.org, November/December.

13  Blumenthal, Max (2010) *Republican Gomorrah: Inside the Movement that Shattered the Party* (New York: Nation Books).

14  Mair, Ruling the Void?

15  Fenby, Jonathan (2015) What the West Should Know about Xi Jinping, China's Most Powerful Leader since Mao, www.newstatesman.com, 23 June.

16  *The Guardian* (2010) US Embassy Cables: French Tell US Britain Is Ready to Abandon Trident, www.theguardian.com, 8 December.

17  Bukovansky, Mlada et al. (2012), Nuclear Proliferation, in *Special Responsibilities: Global Problems and American Power* (Cambridge: Cambridge University Press), 81–121.

18  Acton, James M. (2011) *Deterrence during Disarmament: Deep Nuclear Reductions and International Security* (London: International Institute for Strategic Studies), 23.

19  Acton, James M. (2012) Bombs Away, Being Realistic about Deep Nuclear Reductions, *Washington Quarterly*, 35(2), 38–39.

20  Acton, *Deterrence during Disarmament*; Chalmers, Malcolm (2012) *Less Is Better: Nuclear Restraint at Low Numbers* (Abingdon: Routledge Journals); Deep Cuts Commission Report (2014) *First Report of the Deep Cuts Commission— Preparing for Deep Cuts: Options for Enhancing Euro-Atlantic and International Security* (Hamburg: Institute for Peace Research and Security Policy).

21  World Public Opinion (2008) Publics around the World Favor International Agreement to Eliminate All Nuclear Weapons, www.worldpublicopinion.org, 9 December.

22  Wittner, Lawrence (2010) Where Is the Nuclear Abolition Movement Today?, *Disarmament Forum*, 4 (Geneva: UNIDIR), 7.

23  Sample, Ian (2017) Treaty Banning Nuclear Weapons Approved at UN, www. theguardian.com, 7 July.

24  Interviewees: C, E, L, M.

25  Borrie, John and Caughley, Tim eds. (2013) *Viewing Nuclear Weapons through a Humanitarian Lens* (Geneva: UNIDIR).

26  Sagan, Scott (2010) Shared Responsibilities for Nuclear Disarmament, in Sagan, Scott, *Shared Responsibilities for Nuclear Disarmament: A Global Debate* (Cambridge, MA: American Academy of Arts and Sciences), 7.

27  Dhanapala, Jayantha (2010) Common Responsibilities in the NPT—Shared or Asymmetrical?, in Sagan, *Shared Responsibilities*, 22, 32.

28  United Nations (2014) *United Nations Framework on Climate Change*, http://unfccc.int/2860.php.

29  Pauw, P., Bauer, S., Richerzhagen, C., Brandi, C., and Schmole, H. (2014) *Different Perspectives on Differentiated Responsibilities A State-of-the-Art Review of the Notion of Common But Differentiated Responsibilities in International Negotiations* (Bonn: German Development Institute), 4.

30  United Nations (1992), *Rio Declaration on Environment and Development*, www.un.org.

31  Lusiani, Niko and Muchhala, Bhumika (2015) *Universal Rights, Differentiated Responsibilities: Safeguarding Human Rights beyond Borders to Achieve the Sustainable Development Goals*, www.cesr.org, 2.

32  *Bulletin of the Atomic Scientists* (2017) It Is Now Two and a Half Minutes to Midnight, http://thebulletin.org, January.

33  Rees, Martin (2004) *Our Final Century* (London: Arrow); Ord, Toby (2020) *The Precipice: Existential Risk and the Future of Humanity* (New York: Hachette Books).

34  Ripple, William J. et al. (2017) World Scientists' Warning to Humanity: A Second Notice, *BioScience*, 67(12), December.

35  IPCC (2018) *Global Warming of 1.5°C,* www.ipcc.ch, 10–12.

36  Kamal, Baher (2017) Climate Migrants Might Reach One Billion by 2050, http://www.ipsnews.net, 21 August; Kraan, Caroline M. and Mach, Katharine J. (2019), What's the Relationship between Climate and Conflict?, https://socialsciences.nature.com, 11 July.

37  UK MOD (2015) *Strategic Trends Programme: Future Operating Environment 2035*, https://assets.publishing.service.gov.uk, 4.

38  Klare, Michael (2020) Can the Military's Serious Take on Climate Change Win over More Hearts and Minds?, www.greenbiz.com, 4 March.

39  King, Wendell Christopher (2014), *Climate Change: Implications for Defence: Key Findings from the Intergovernmental Panel on Climate Change*, http://gmaccc.org/, 4.

40  Johnson, Keith (2019) Are India and Pakistan on the Verge of a Water War?, https://foreignpolicy.com, 25 February.

41  Hamza, Shaban (2014) Climate Change Is an Opportunity to Dramatically Reinvent the Economy, www.theatlantic.com, 19 September.

42  Green New Deal Group (2020), https://greennewdealgroup.org.

43  Monbiot, George (2020) Let's Nuke the Virus, www.monbiot.com, 9 April; Olivier, Indigo (2020) Time to Redirect the Military Budget to Public Health, http://inthesetimes.com, 8 April; Stark, Alexandra (2020) COVID-19 Is this Generation's 9/11. Let's Make Sure We Apply the Right Lessons, https://www.newamerica.org, 23 April.

44  IPSOS Global Advisor (2020) *How Does the World View Climate Change and Covid-19?*, www.ipsos.com, 6.

# Conclusion

This study has explored the politics of nuclear disarmament in order to develop an original and compelling analysis of the obstacles to and opportunities for eliminating nuclear weapons in and between the five NWS. Chapter 1 investigated the different ways that nuclear disarmament has been conceptualised and justified. On a technical level it was found that nuclear disarmament should be seen as a learning process, which would build on existing knowledge across several phases, including: i) restraint, ii) elimination, iii) maintaining zero. As an NWS passes through these phases towards FNWS status, so the requirements of disarmament will become more onerous and the meaning of elimination clearer, including in terms of the resources and degree of intrusion necessary to monitor and verify the dismantlement of nuclear capabilities. In addition to verifiability and transparency, such disarmament actions will need to be irreversible, if they are to be implemented in line with NPT obligations, with higher levels of irreversibility making rearmament more difficult.

As the disarmament process unfolds, and depending on the number of possessor states involved, the degree of international cooperation and type of political institutions necessary to achieve and lock in nuclear zero—including the extent to which power needs to be centralised at the international level given the dual-use nature of the nuclear fuel cycle—will become increasingly clear. For example, it is highly likely that the international processes that will facilitate, and evolve alongside, the legal and technical instruments and knowledge required for a nuclear disarmament process involving China, Russia and the US will require close and sustained political and security cooperation between these states to ensure that progress is not derailed. In other words, the more stable and peaceful relations between the major powers are, the better the prospects will be for them to diminish their reliance on nuclear weapons and progressively implement disarmament action.

Chapter 2 then turned to political theory, exploring existing approaches to nuclear possession and disarmament in mainstream and realist thought and scholarly critiques of realism. Having reviewed a range of prominent works regarding the meaning of nuclear weapons—most of which primarily, if not exclusively, focus on the international security implications of the bomb—I identified several significant gaps and problems. For example, according to scholars such as Mearsheimer, Glaser, Waltz and Schelling—who strongly criticise or oppose nuclear disarmament concepts—a disarmed world would be more unstable than a nuclear-armed world, may increase the probability of nuclear conflict and is, in any case, improbable given the various obstacles to international cooperation. Others also argue that nuclear disarmament involving the major powers would generate security dilemmas, leading, for example, to conventional arms races given the difficulty of differentiating between defensive and offensive weaponry and the fact that states cannot be certain about each other's intentions.

Whilst these concerns and objections have some merit, they are not sufficiently powerful to demolish the case for nuclear disarmament. This is first because, as even those mainstream and realist scholars who strongly support nuclear possession—including Mearsheimer—note, it is not possible to be certain of the extent to which nuclear weapons contributed to preventing great-power conflict during the Cold War. Moreover, others, such as Jervis, go further and question whether nuclear weapons have been beneficial at all for strategic stability, given the possibility of cataclysmic nuclear war. The fact that those objecting to nuclear disarmament on the grounds that it would lead to unacceptable levels of instability cannot prove that a nuclear-armed world is more stable and less likely to create nuclear conflict should lead us to question the veracity of their argumentation.

When we do so, as in the case of Mearsheimer and others, we find that the 'stability' they refer to principally concerns the narrow aims and interests of the US and its allies, rather than any universal conception of security. Yet US national security goals defined by elite decision-makers are, alone, not a legitimate basis on which to make judgements concerning whether nuclear disarmament is desirable or realisable. Rather, because this subject is of utmost concern to the security of the entire US populace—and, for that matter, the world—we need to find an alternate, legitimate means of making judgements and decisions on nuclear issues.

I argue that a new approach is also necessary because the advent of nuclear weapons had implications for possessor states that are not sufficiently explored in the existing mainstream and realist literature. This primarily concerns the domestic political impact of states acquiring and developing the bomb and thus what eliminating nuclear weapons will entail on the domestic economic, social and political fronts. The alternative approach proposed, which I termed *institutional democratisation,* therefore focuses on domestic explanations of nuclear possession and the need for a normative theory of nuclear disarmament which identifies the appropriate domestic political reforms required if NWS are to completely eliminate their nuclear arsenals.

Furthermore, *institutional democratisation* engages with the problems raised by mainstream and realist thought concerning how anarchy may be mitigated or overcome, where the political will to sustain international cooperation pursuant to nuclear disarmament in and between NWS might originate, and how future international institutions supportive of disarmament and a NWFW may avoid descending into an illegitimate hierarchy or tyranny. For example, *institutional democratisation* argues that, if citizens of an NWS are empowered to engage with, participate in and direct political decision-making in key areas, this will increase that NWS's level of international cooperation pursuant to the creation of just and equitable regimes and institutions.

Increasing civilian and democratic influence and control over NWS's defence and foreign policy, as appropriate to the circumstances of each, would also likely help moderate and restrain these state's international behaviour. Currently, whilst there are, albeit to different degrees, significant constituencies within NWS supportive of nuclear arms control and disarmament action, they are at best, only weakly represented in NWS's nuclear policies. This is primarily because of the dominant influence of military-security elites over decision-making, inadequate processes of democratic oversight and control, the relatively low salience of nuclear issues for citizens and the lack of institutional focus for alternative, non-nuclear policies that variously exist within NWS.

The potential for international cooperation pursuant to nuclear disarmament (most notably in the US) thus exists in NWS, but is not being realised, often because of domestic bureaucratic continuity and political obstruction, driven by decision-makers' wider strategic goals. Moreover, the secrecy and misinformation surrounding nuclear issues, often purposefully fed by pro-nuclear governments and their supporters, has led to the citizens of NWS being misled regarding the actual costs and risks of nuclear possession. Educating and informing the public about the meaning of maintaining and modernising the bomb, and what alternative options exist, are thus crucially important steps to build support for nuclear disarmament as part of a process of *institutional democratisation*.

Part II of this study focused on showing how *institutional democratisation* adds value to mainstream and realist explanations specifically concerning the causes and consequences of NWS nuclear possession and disarmament. In order to do this, Chapters 3 to 7 provided in-depth case studies of the nuclear weapons systems of each of the five NWS in order of their acquiring the bomb: US, Russia, UK, France and China. Over these case studies, it was found that, even in the cases where the security model does a good job of explaining the causes and consequences of NWS nuclear acquisition and development by focusing on state's perception of external threats to their sovereignty and national security, it still only provides a partial picture. The key point here is that, without a full grasp of why NWS sought—and then continued to possess and develop—the bomb, it is not possible to identify the politics of nuclear disarmament for each NWS.

For example, by focusing on the national security aspects of nuclear acquisition, realist thought is useful in highlighting the external threats to the homeland which motivated Russia and China, in particular, to seek and maintain possession of the bomb. Moreover, in terms of the transformation in international security potentially required for nuclear disarmament, realism reminds us that it is not reasonable to argue that NWS disarmament or a NWFW will be the same world as today, just without nuclear weapons. This is because US military power in Europe and East Asia is perceived as a threat to Russian and Chinese sovereignty and independence and drives decision-makers in Moscow and Beijing to maintain their reliance on nuclear deterrence to ensure regime survival, protecting their interests at home and in their near abroad. The fact remains, however, that there are clear problems with mainstream and realist approaches to nuclear possession and disarmament that must be recognised. For example, these approaches do not fully account for or consider two key aspects of nuclear politics requiring theories of change: i) each NWS's continued possession and development of nuclear weapons over time, particularly following the end of the Cold War, ii) NWS nuclear disarmament. These gaps in the orthodox literature are problematic for several reasons.

First, despite the existential threats facing today's world, where nuclear arms have a high and increasing salience, mainstream and realist approaches are not greatly interested in imagining or even discussing the causes and beneficial consequences of disarmament. As noted above, this is because such works generally embrace the nuclear revolution, principally based on the belief that MAD prevented superpower conflict during the Cold War. However, to explain the NWS's continued refusal to eliminate their nuclear arsenals following the end of the Cold War we need to look more closely at nuclear weapons decision-making in each NWS and consider drivers of nuclear possession not accounted for by the security model.

When we do so, we find that elite decision-makers in the three Western NWS—France, the UK and US—but also, to an extent, Russia, believe they have a 'special responsibility' to manage global order, a role which they maintained beyond the fall of the Soviet Union, and which they also believe confers 'special rights', including for foreign military interventions and indefinite nuclear possession. Importantly, the ability of each NWS to realise their strategic goals must be differentiated here, based on their available resources and capabilities, so that the Western NWS are and were collectively far more powerful than Russia and therefore have greater freedom to act. Notably, whilst China has historically not been an NWS 'in the Western sense' it may, in future, decide to move in the direction of becoming a global military power and configure its nuclear arsenal accordingly.

It may therefore be optimistic yet still reasonable to believe that existing elites and institutions in NWS will, as per the *guardianship* model, gradually enact more enlightened policies of arms control, non-proliferation, restraint and other measures supportive of disarmament, for reasons of 'strategic stability'—as has sometimes been done in the past following international conflict or crises—or in response to cost pressures or technological developments. What is not reasonable or realistic is to believe that business as usual can continue in perpetuity without a more dangerous global nuclear order emerging or that existing NWS decision-making elites have the will or vision to independently make significant strides towards nuclear abolition, nationally and internationally.

Instead, permanent bureaucracies in NWS are optimising the effectiveness of their nuclear arsenals by producing ever more advanced, threatening and potentially usable weaponry. Technological developments in conventional arms, such as precision-guided munitions, as well as other emerging cyber and AI capabilities, also pose dangers of military entanglement and unintended escalation involving China, Russia and the US—which could potentially lead to nuclear war. At the same time, such complex modernisation efforts are increasingly costly and difficult for NWS—particularly for the US and Russia—requiring advanced industrial skills and significant resources, which may act as a brake on the speed with which these systems can be successfully deployed, presenting an opportunity for appropriate arms control and disarmament regimes to be agreed between the major powers.

Given the limitations of mainstream and realist thought identified above, the value of *institutional democratisation* lies in its ability to fill in the gaps regarding why states acquire the bomb, explain the particular ways in which states develop the bomb and what needs to change politically in and between NWS if disarmament is to advance. This is done by bringing in to our analysis a range of important domestic factors which variously influence and determine nuclear policy for each NWS, including: the weight of national history resting on decision-makers' shoulders—whether victorious or traumatic, security-seeking or expansionist; the influence of bureaucratic continuity as well as scientific and technological prestige; levels of democratic oversight, transparency and accountability; the extent of public activism, engagement and debate; as well as the other domestic economic, political and strategic benefits nuclear possession and development brings to those elite actors in control of decision-making.

In addition, *institutional democratisation* outlines how NWS may eliminate the bomb, by providing a viable normative theory of domestic political change. For example, if the NWS are to complete the elimination of their nuclear arsenals, then this will likely require the formation of popular social movements to transform domestic power structures, as appropriate to the circumstances of each NWS, including establishing and reviving democratic institutions in order to redistribute power and reduce the influence of military and security elites over decision-making. These disarmament requirements apply, crucially, to the US, given its singular influence, power and global reach. China, meanwhile, is a lesser included case, primarily given its history of nuclear restraint, authoritarian political system and perceived need to deter US threats, so that it would require sustained improvements to its regional security if it were to disarm.

After the explanatory power of *institutional democratisation* regarding the politics of NWS nuclear possession and disarmament was demonstrated in Chapters 3 to 7, Chapter 8 then moved on to consider the practical political implications of *institutional democratisation,* to illustrate its value as both a critical theory and problem-solving tool. With regards to the specific ideas and processes supportive of NWS *institutional democratisation* that could be taken forward, three areas were highlighted. First, the need for political analysts to develop measurements of authoritarianism and democracy in NWS reflecting the costs and risks of nuclear possession. This is necessary in order to both better appreciate the benefits of nuclear disarmament to civil liberties and more accurately define a state's regime type, including whether a democracy is more popular or elitist. Second, the need for NWS governments to implement progressive policies ensuring that nuclear weapons are subject to democratic, transparent and accountable processes. Third, the need for civil society to improve public understanding of nuclear weapons and disarmament in NWS, where possible, as a means of developing greater popular engagement with and participation in decision-making on national security issues.

Based on this study's claim that each NWS has a dual responsibility for nuclear disarmament, covering action both at a national level and an international level that supports other NWS taking action to disarm, the third section of Chapter 8 then considered the current health of pro-disarmament movements in each NWS and globally to assess the potential for action supportive of *institutional democratisation* to be realised in the near term. For example, in order to develop a stronger analysis of the obstacles to and opportunities for eliminating nuclear weapons today, it was proposed that advocacy and campaign groups working in this area could produce strategies to advance the cause of disarmament based on a differentiation between states (including NWS) in terms of their responsibilities for disarmament. Such an analysis would need to take into account and weigh a state's overall military capabilities and postures, both conventional and nuclear. This would potentially be a powerful tool for disarmament advocates in visualising and communicating the obstacles to and opportunities for a NWFW, as well as how possessor states can reduce their nuclear arsenals to converge on zero.

Given the US's exceptional global power and the fact that its behaviour has a decisive influence on all nations' nuclear weapons decision-making, the case for supporters of nuclear disarmament today to focus primarily on the US is thus a strong one. In addition, because all NWS governments need to be persuaded or pressured to realise their dual nuclear disarmament obligations, pro-disarmament political strategists, including those campaigning for possessor states to support a TPNW, need to think carefully about where they devote resources and how they frame the issue so that Western-centric perspectives do not predominate. In considering this we must recognise that some, perhaps the majority, of those who spend most time working for NWS nuclear disarmament do so from within the NWS of which they are citizens—principally the UK and US—but also across Europe, Japan, Australasia and NAM members.

The reason for doing so seems clear and practical, individuals and NGOs are far more likely to be able to influence decision-makers if they are citizens of that nation through processes of democratic accountability. Clearly, prevailing political conditions, for example, levels of democracy and freedom in each NWS, will affect how easy or hard it is to do this. Furthermore, citizens and groups working in nations which are relatively more liberal, democratic and secure, in particular the UK and US, with France a lesser included case, have greater access to resources and channels of influence than those living under less democratic or free regimes such as China and Russia.

At the same time, it is important to recognise that the lack of an effective review process in Western, formally democratic NWS and the lack of public accountability of those shaping nuclear decisions, offers more similarities than differences with the process of nuclear weapons decision-making in Beijing and Moscow's more authoritarian regimes. This raises the question of how influential individuals and groups outside of the upper echelons of the bureaucratic, military and political elites can be, even within relatively free societies with at least formally democratic institutions, such as the US.

Research conducted in several democracies, including in Europe, Japan and the US, has found that where anti-nuclear weapons public opinion, protest and civil society activism exists, it has exerted an influence on the degree and type of action taken by governments. However, it must also be recognised that, despite the rising risks of major power conflict escalating to nuclear war, the deep secrecy and lack of visibility concerning nuclear weapons systems has acted as a strong brake on the level of public engagement in and action on these matters. This fact helps explain why, despite nuclear disarmament advocates in the West campaigning for decades on this issue, nuclear weapons systems—and the institutions that produce and maintain them—have remained so resilient in the face of external pressure, and why, as several leading disarmament activists have attested, there has been an apparent turning away by the public from nuclear issues, despite pro-disarmament views remaining widespread.

Despite the significant challenges facing pro-disarmament groups, there is great potential today for them to link up with campaigners working in related fields, such as conflict, climate change, poverty and development, to develop a radical critique of power and propose short-, medium- and long-term measures supportive of a transition to more equitable and just societies. Indeed, it is likely that movements with a wider social base and goals, which can draw on greater public support and resources, will be essential if economic elite domination and cultures of militarism are to be replaced with democratic, participatory economic and political systems capable of demilitarising NWS.

In terms of the future of nuclear disarmament, there is a strong case to be made that the politics of nuclear weapons and climate change in particular should be seen as the politics of human survival. Without the emergence of strong and diverse popular movements, particularly in the US and its allies, to make the costs and risks of nuclear possession visible as a public concern, increase pressure for arms control and disarmament action, and eventually preside over state institutions to control the power of nuclear weapons pending their elimination, the prospects for scientific societies to continue existing and provide decent lives for people over the long term may fast diminish.

Whilst the current state of the world may foster pessimism in observers, there are several historical precedents proving that social transformations of a comparable scale and significance to those which nuclear disarmament entails are possible, including the abolition of slavery, the end of the Cold War, the overthrow of apartheid, de-colonisation and other advances in national independence, democracy and civil rights. The type of emancipatory politics that inspired the popular struggles which achieved these victories for social justice will also enable NWS to achieve nuclear disarmament. To ensure humanity's common prosperity, security and survival, citizens in NWS and from across the world will therefore need to come together to ensure that democratic principles guide nuclear possessor's decision-making at a fundamental level, a process that necessitates *institutional democratisation*.

# Index